Localization and Mapping of Autonomous Mobile Robots

Localization and mapping play a critical role in the autonomous task execution of mobile robots. This book covers the theoretical and technological aspects of robot localization and mapping, including visual localization and mapping, visual relocalization, LiDAR localization and mapping, and place recognition.

It provides the theoretical foundations of robot localization and mapping. It employs both traditional methods, such as geometry-based visual localization, and state-of-the-art deep learning techniques that improve robot perception. The authors also address LiDAR-based localization, exploring techniques to improve both efficiency and accuracy when processing dense point clouds. Key topics include visual localization using deep features, integration of visual solutions under ROS-based software architecture, and distribution-based LiDAR localization.

This book will be of great interest to students and professionals in the fields of robotics and artificial intelligence. It will also be an excellent reference for engineers and technicians involved in the development of robot localization.

Junzhi Yu is a Professor in the Department of Advanced Manufacturing and Robotics, College of Engineering, Peking University, China, and also a guest researcher at the Institute of Automation, Chinese Academy of Sciences. His research interests include intelligent robots, motion control, and intelligent mechatronic systems. He has authored or co-authored seven monographs and published more than 200 science citation index papers in prestigious robotics and automation-related journals. He has successively been listed among the Most Cited Researchers in China between 2014 and 2023. He is a fellow of the Institute of Electrical and Electronics Engineers.

Zhiqiang Cao is currently a Professor at the State Key Laboratory of Multimodal Artificial Intelligence Systems from the Institute of Automation, Chinese Academy of Sciences. His current research interests include service robots and intelligent robots.

Peiyu Guan is currently an Assistant Professor at the State Key Laboratory of Multimodal Artificial Intelligence Systems from the Institute of Automation, Chinese Academy of Sciences. Her research interests include service robots, visual localization and mapping, and computer vision.

Chengpeng Wang is at the State Key Laboratory of Multimodal Artificial Intelligence Systems from the Institute of Automation, Chinese Academy of Sciences. His research interests include intelligent robots, robot localization, and navigation.

Localization and Mapping of Autonomous Mobile Robots

Junzhi Yu, Zhiqiang Cao, Peiyu Guan,
and Chengpeng Wang

CRC Press
Taylor & Francis Group
Boca Raton London New York

CRC Press is an imprint of the
Taylor & Francis Group, an **informa** business

Designed cover image: © Junzhi Yu, Chengpeng Wang, Zhiqiang Cao, and Peiyu Guan

This book is published with financial support from the National Natural Science Foundation of China under Grant 62303456 and Grant T2121002, and the Beijing Natural Science Foundation under Grant L233030 and Grant 2022MQ05.

First edition published 2026
by CRC Press
2385 NW Executive Center Drive, Suite 320, Boca Raton FL 33431

and by CRC Press
4 Park Square, Milton Park, Abingdon, Oxon, OX14 4RN

CRC Press is an imprint of Taylor & Francis Group, LLC

ISBN: 978-1-032-91704-7 (hbk)
ISBN: 978-1-041-08366-5 (pbk)
ISBN: 978-1-003-64363-0 (ebk)

DOI: 10.1201/9781003643630

Typeset in Minion
by codeMantra

Contents

Figures and Tables

CHAPTER 1

Introduction

1.1 RESEARCH BACKGROUND

As a representative of intelligent equipment, robots integrate the technologies including mechanical engineering, electrical engineering, information processing, control, and artificial intelligence, with widespread applications in manufacturing industries, medical health, urban search and rescue, household services, etc. Nowadays, robots have been developed rapidly, making them an important force to promote a new quality productivity. To execute various tasks for autonomous robots, the localization and mapping are indispensable prerequisites, where the pose of the robot is determined by processing the information provided by on-board sensors. The household services, autonomous cleaning, unmanned delivery, etc. all depend on the localization technology for path planning [1–4]. The localization is also applied to auxiliary construction of high-definition maps in autonomous driving. Besides, it can provide support to determine the pose of wearer in the field of augmented reality or virtual reality. The research on localization and mapping is important with great value.

The sensor is the source of data for localization and mapping, and the commonly used sensors include Beidou navigation satellite system (BDS), global positioning system (GPS), ultra-wideband (UWB), camera, light detection and ranging (LiDAR), and inertial measurement unit (IMU). The principle of BDS/GPS positioning is to determine the absolute position of a receiver according to the time it takes for signals from multiple artificial satellites to reach it. A limitation of such positioning is that it is unsuitable to operate in closed environments such as building interior space or underground passages. UWB positioning utilizes short-duration, low-energy radio pulses over a wide frequency range, and the position of a UWB tag is determined with the help of multiple anchors with known positions. It is an effective indoor localization manner; however, the obstacles in the environment often interfere with the transmission of signals, decreasing the accuracy of localization. IMU measures the carrier's motion including angular velocity and acceleration, and then the pose is estimated after integration. Such processing will lead to the accumulation of localization errors. It is not a good choice to execute localization only using IMU, and it is suggested to serve as an auxiliary role. Compared to the aforementioned manners, visual and LiDAR localizations are more common, flexible, and popular, where the carrier pose is estimated based on the association of continuous observations. The visual sensors enjoy the advantages in size, weight, and price with abundant color and texture information, whereas the LiDAR sensor can provide accurate measurements with a wide field of view and robustness to illumination variations. Visual/LiDAR localization and mapping are the two most mainstream branches, which are the focus of this book. It is worth mentioning that localization and mapping complement each other, which are usually implemented under the framework of simultaneous localization and mapping (SLAM). A whole SLAM procedure consists of an odometry, loop closing and global pose optimization, and mapping. The odometry constantly estimates the sensor pose of the current frame. Loop closing is used to judge whether the robot has passed through its previous location. If a loop is detected, the sensor poses at different frames are jointly optimized to promote the global consistency of the trajectory. With these optimized sensor poses, a global map related to a specific task can be easily obtained. It is seen that odometry is a fundamental and core localization unit of SLAM and it can run alone. Notice that the pose provided by SLAM is the one relative to the initial frame, and it will vary with different starting positions. To ensure a unified global pose, a preferable way is to construct an offline global map through SLAM. Furthermore, it is not always sufficient to localize the robot by using the online map, especially during the long-term task execution. The application of offline global map becomes meaningful to relocalize the robot in cases such as SLAM tracking failure. SLAM solves the localization with unknown environment maps while relocalization can be seen as the localization under a known map, and they both belong to the scope of localization and mapping. Next, this chapter summarizes two aspects of research status: visual and LiDAR localization and mapping.

1.2 REVIEW OF VISUAL LOCALIZATION AND MAPPING

For visual localization and mapping, we mainly review visual SLAM for an unknown environment map and visual relocalization for a known map. Visual SLAM typically associates pixels from the current image with those from the previous image or point cloud derived from the local map, and pose estimation is then performed by solving for the camera transformation that best aligns these correspondences. Based on the way of data association, visual SLAM can be broadly categorized into direct method [5,6] and feature-based method [7–10]. The former directly projects points from the previous frame onto the current frame using a pose hypothesis for data association, calculates the photometric error between corresponding points in the two frames, and optimizes the camera pose iteratively by minimizing this error while refining data associations. In contrast, feature-based method extracts key features such as keypoints and lines from the images, matches features between different frames through descriptor distances or optical flow tracking for data association, and estimates the camera pose based on these well-associated features. Overall, feature-based methods focus on computational efficiency and robustness in well-textured environments, while direct methods attempt to improve accuracy in more challenging scenarios by using raw image data at the cost of higher computational complexity.

Visual relocalization aims to restore the global pose of robot by combining the information of the current image and known map, where the map can be represented using explicit three-dimensional (3D) structure or implicit model with amounts of parameters. Therefore, relocalization methods can be divided into explicit map-based [11,12] and implicit map-based [13,14] methods. Explicit map-based relocalization relies on pre-built 3D environment map that stores keyframes and detailed geometric information, such as 3D points and their corresponding 2D features from images. The robot pose is then solved by local feature matching between the current frame and the pre-built map. In contrast, implicit map-based relocalization avoids the explicit local feature matching and expresses the map as a trained model that stores scene information implicitly, which can be further classified into scene coordinate regression-based method [13] and global pose regression-based method [14]. Scene coordinate regression-based method first uses random forests or neural networks to predict the 3D world coordinates corresponding to each 2D pixel in the input image, and the resulting 2D–3D correspondences are then input into a RANSAC-based PnP solver to obtain the camera's 6D pose [15]. Global pose regression-based method regresses the camera pose directly via a model trained on the images of environment map.

1.2.1 LSD-SLAM

Engel et al. proposed LSD-SLAM [5], which was a direct monocular SLAM method capable of building large-scale, consistent maps of the environment. As shown in Figure 1.1, it consists of three major components: tracking, depth map estimation, and map optimization. The tracking component continuously tracks new camera images. Using the pose of the previous frame as initialization, the input image is aligned to the current keyframe to estimate the SE(3) transformation of the current frame with respect to the current keyframe by minimizing the variance-normalized photometric error. Next, the current frame is sent to the depth map estimation module. A keyframe selection strategy based on weighted distance is designed to decide whether the current frame should be selected as a keyframe.

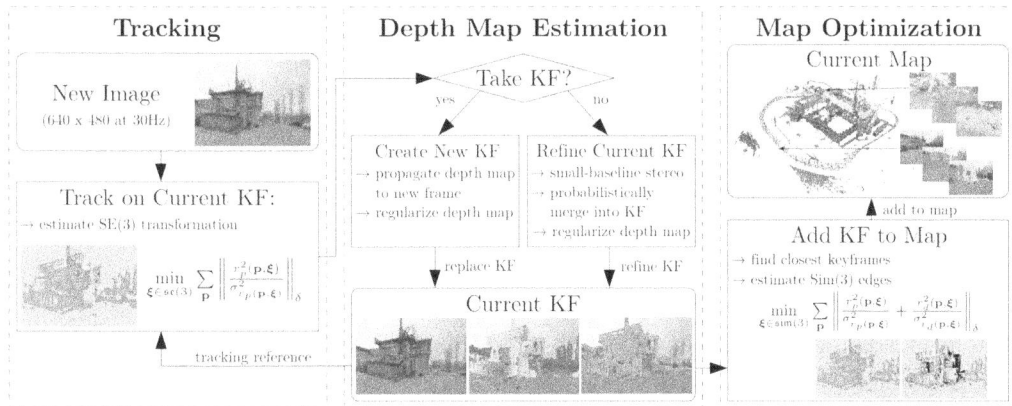

FIGURE 1.1   The framework of LSD-SLAM [5].

If chosen, a new keyframe is created and its depth map is initialized by projecting points from the previous keyframe into it. This is followed by spatial regularization and outlier removal. The depth map is then scaled to have a mean inverse depth of 1. The resulting new keyframe replaces the previous keyframe for tracking subsequent new frames. Otherwise, the input frame is used to refine the depth map of the current keyframe. Specifically, small-baseline stereo comparisons are performed to obtain the new stereo depth measurement, which is probabilistically merged into the depth map of the current keyframe [16]. After regularizing the merged depth map, the current keyframe is refined. In the map optimization module, the closest keyframes and loop closures are found for the current keyframe, and the corresponding Sim(3) transformations are estimated by minimizing the sum of photometric and depth residuals. Leveraging these Sim(3) constraints, the map is continuously optimized using pose-graph optimization.

1.2.2 Direct Sparse Odometry (DSO)

Engel et al. [6] proposed a direct and sparse monocular visual odometry, which was based on a direct probabilistic model that minimizes photometric error over a window of recent frames. Considering that the direct approach models the full image formation process down to pixel intensities, it is beneficial to utilize a more precise sensor model. Therefore, DSO formulates a photometrically calibrated model including geometric and photometric camera calibration for image formation. The geometric camera calibration part adopts the pinhole camera model to project a 3D point onto the 2D image, while the photometric camera calibration part combines the non-linear response function, lens attenuation (vignetting), and exposure time in [17] to map real-world energy received by a pixel on the sensor (irradiance) to the respective intensity value. On this basis, a new photometric error considering the camera intrinsics is derived. Besides, the existing dense or semi-dense direct methods typically exploit the connectedness of the used image region to formulate a geometric smoothness prior, which impedes the real-time performance of the methods. To deal with this problem, DSO samples pixels from all image regions with intensity gradients, such as edges or smooth variations on featureless walls, and evaluates the photometric error for each sampled point over a small neighborhood of pixels to constrain the overall problem effectively. The full photometric error over all frames and points is further used in a sliding window to jointly optimize all model parameters, including camera poses, camera intrinsics, and geometry parameters (inverse depth values). Figure 1.2 displays the built maps for three scenes along with their corresponding depth maps, demonstrating that DSO is capable of tracking through scenes with very little texture.

1.2.3 ORB-SLAM Family

1.2.3.1 ORB-SLAM

Raul Mur-Artal et al. proposed ORB-SLAM [7] in 2015, which is one of the most classical feature-based monocular visual SLAM algorithms. As shown in Figure 1.3, it integrates three threads that run in parallel: tracking, local mapping, and loop closing. The tracking thread is in charge of localizing the camera with every frame and deciding when to insert

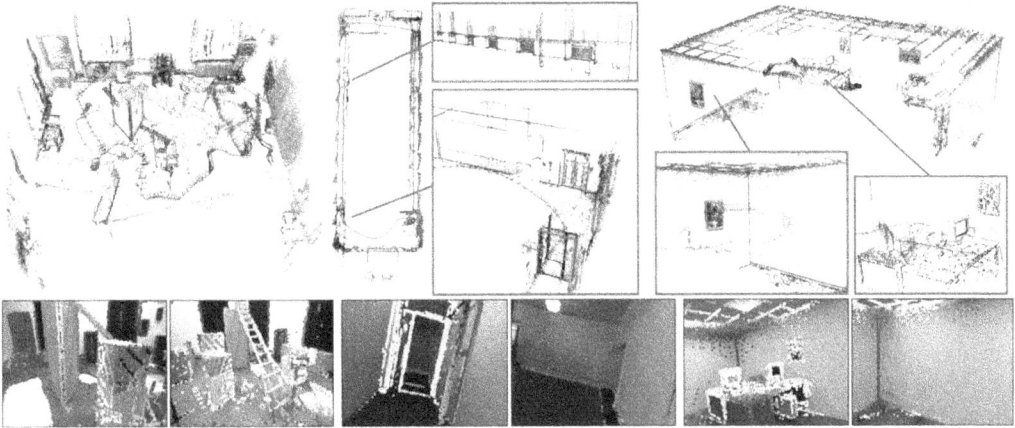

FIGURE 1.2   The visualization of DSO runtime results [6].

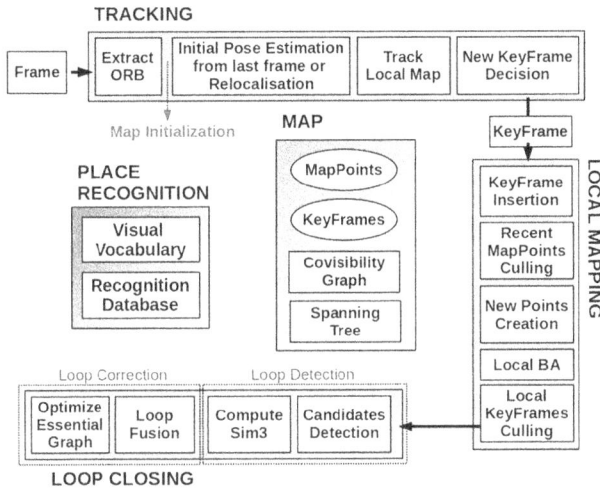

FIGURE 1.3   The framework of ORB-SLAM [7].

a new keyframe. Oriented FAST and rotated BRIEF (ORB) feature points [18] are first extracted from the input monocular images. Then these feature points are matched with those in previous frames. On this basis, camera poses are estimated through motion-only bundle adjustment. If the above tracking fails, the relocalization is activated based on the place recognition module in Figure 1.3. The place recognition module consists of visual vocabulary and recognition database. Based on DBoW2 [19], the visual vocabulary is created offline with the ORB descriptors extracted from a large set of images. The recognition database stores the related keyframes for each word in this vocabulary. Once there is an initial estimation of the camera pose and feature matchings, the local map is projected into the current frame to find more matching correspondences, and then the camera pose is optimized again with all the matched points. Those frames that satisfy set conditions are decided as new keyframes. After that, keyframes are fed into local mapping thread. The local mapping mainly processes new keyframes and performs local bundle adjustment

(BA) to achieve an optimal reconstruction in the surroundings of the camera pose. First, keyframes are inserted to update the covisibility graph and spanning tree. Herein, covisibility graph is an undirected weighted graph whose nodes are keyframes and whose edges represent the covisible relationship between two keyframes. Next, an exigent point culling policy is utilized to ensure that only high-quality map points are left. Then, new map points are created through triangulation of matched 2D feature points to provide more constraints for optimization. Subsequently, local BA is performed to optimize the currently processed keyframe and those keyframes connected to it in the covisibility graph. In addition, some redundant local keyframes are culled to ensure their number will not grow unbounded. The loop closing thread searches for loops with every new keyframe. The loop candidates are detected by computing the similarity of the bag-of-words vector of the current keyframe with those of the keyframes in covisibility graph. For each of the detected candidates, we compute the Sim3 transformation to estimate the relative poses and scale factor. According to the relative transformation, loop fusion is executed to align both sides of the loop and fuse the duplicated points. To make sure efficient pose optimization, ORB-SLAM retains all the nodes but fewer edges from the covisibility graph to construct an essential graph. Eventually, essential graph optimization is performed to obtain the more precise and consistent map.

1.2.3.2 ORB-SLAM2

Mur-Artal et al. proposed ORB-SLAM2 [8] in 2017. Built on ORB-SLAM, ORB-SLAM2 is a complete SLAM system for monocular, stereo, and RGB-D cameras. Its framework is shown in Figure 1.4a, which can be seen to be similar to the framework of ORB-SLAM. However, there exist some main differences. Firstly, an input preprocessing step is added in the tracking thread so that the rest of the system is independent of the input sensors. The detailed preprocessing pipeline is illustrated in Figure 1.4b. For rectified stereo image pairs, ORB feature points are extracted and points in the left image are matched with those in the right image. Then the stereo keypoints are generated with the coordinates of

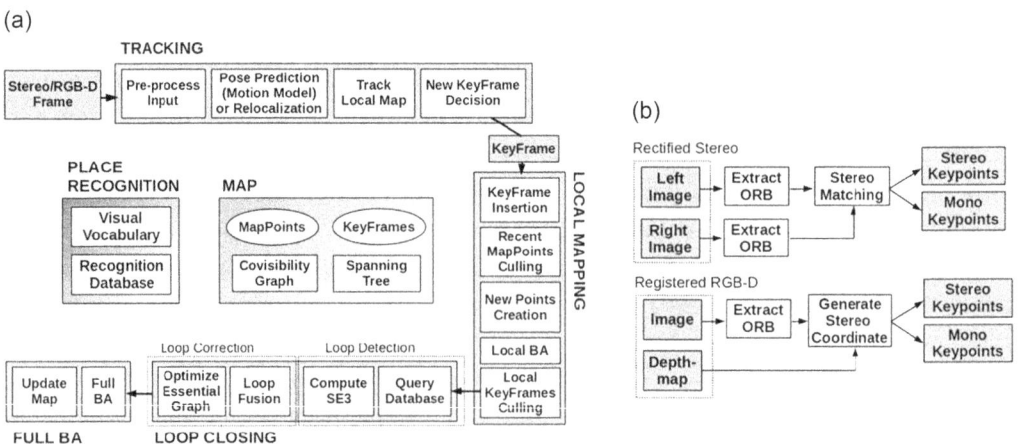

FIGURE 1.4 ORB-SLAM2 [8]. (a) The framework of ORB-SLAM2. (b) Input preprocessing.

the left ORB and the horizontal coordinate of the right match. For a registered RGB-D image, ORB feature points are extracted on the RGB image, and the depth value of each feature point is transformed into a virtual right coordinate to form the stereo keypoints. As a result, features from stereo and RGB-D input are handled equally by the rest of the system. Then, the projection function in BA module is also extended to adapt to both monocular and stereo keypoints. In contrast to the Sim(3) computation employed in ORB-SLAM, ORB-SLAM2 has been refined with the introduction of SE(3) computation within the loop closing thread, specifically tailored for stereo and RGB-D modes. In these modes, the scale factor is inherently determined, as the availability of stereo or depth information renders the scale directly observable. Besides, a full BA process is added after loop correction in a separate thread, which optimizes all keyframes and points to provide a globally consistent map.

1.2.3.3 ORB-SLAM3

Carlos Campos et al. proposed ORB-SLAM3 [9] in 2021 on the basis of ORB-SLAM2. ORB-SLAM3 is the first system able to perform visual, visual-inertial, and multimap SLAM with monocular, stereo, and RGB-D cameras, using pinhole and fisheye lens models. Figure 1.5a displays its main components, similar to those of ORB-SLAM2, but also includes some major innovations. With the introduction of IMU information, ORB-SLAM3 becomes a tightly integrated visual-inertial SLAM system that fully relies on maximum a posteriori (MAP) estimation. Specifically, IMU integration is involved in the tracking thread, which provides motion observation between two consecutive visual frames by pre-integrating the IMU measurements on manifold [20]. On this basis, the inertial residual is constructed and combined with visual residual to formulate a visual-inertial optimization problem for accurate state estimation. This optimization can be adapted for efficiency during motion-only BA in tracking and local BA in mapping. More importantly, it requires good initial estimates for inertial variables (body velocities, gravity direction, and IMU biases) to converge to accurate solutions. Since IMU initialization depends on a reliable initial pose, a fast and accurate IMU initialization method is proposed to be executed after the local visual BA in the local mapping thread. It is stated as a MAP estimation problem and divided into three steps: vision-only MAP estimation, inertial-only MAP estimation, and visual-inertial MAP estimation. To address initialization failure caused by insufficient observability of the inertial parameters during slow motion, an efficient scale refinement strategy is applied within the local mapping thread, focusing solely on optimizing the scale and gravity direction parameters. Different from the single map in ORB-SLAM2, ORB-SLAM3 develops a multimap representation known as Atlas, which consists of a collection of disconnected maps. Within Atlas, the map in which the current frame is localized is called active map, while the other maps are regarded as non-active maps. When tracking is lost, the tracking thread attempts to relocalize the current frame across all maps in the Atlas. If it fails, the active map is marked as non-active and a new active map is created. In loop and map merging thread, if the detected common area belongs to the active map, loop correction is performed; if it belongs to a different map in Atlas, both maps are seamlessly merged into a single one, which becomes the active map. The test result

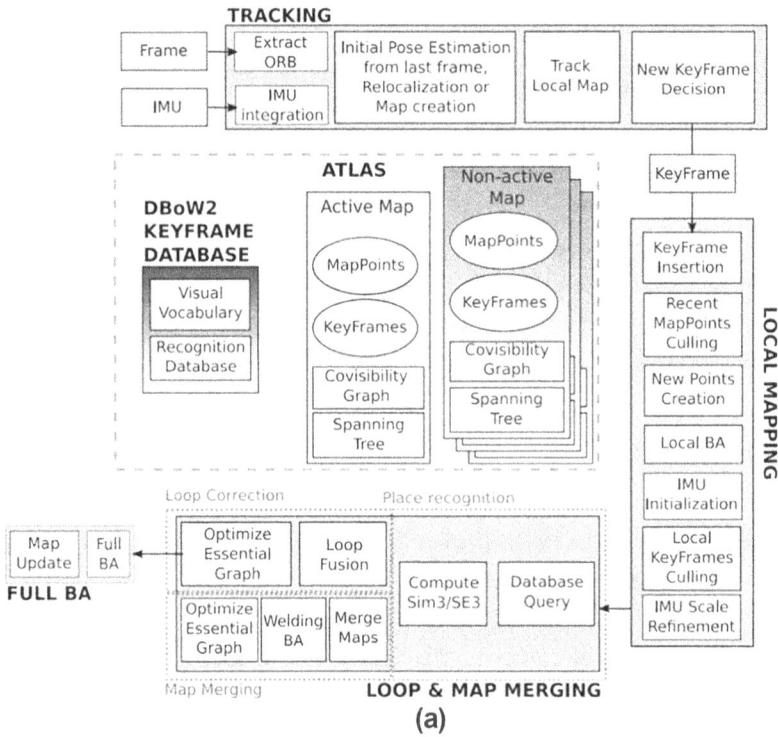

FIGURE 1.5 ORB-SLAM3 [9]. (a) The framework of ORB-SLAM3. (b) Visualization of ORB-SLAM3's runtime results. The left figure illustrates the original image with tracked feature points, while the right one displays the reconstructed map.

of ORB-SLAM3 on EuRoC dataset is visualized in Figure 1.5b, which shows the system's capability to handle challenging real-world scenarios.

1.2.4 PL-SLAM

Pumarola et al. proposed PL-SLAM [10] in 2017, which is a real-time monocular visual SLAM with points and lines. Built upon ORB-SLAM, PL-SLAM uses not only points but also lines as features to construct constraints and estimate poses, thus greatly improving accuracy and robustness in low-textured scenes. As shown in Figure 1.6, PL-SLAM

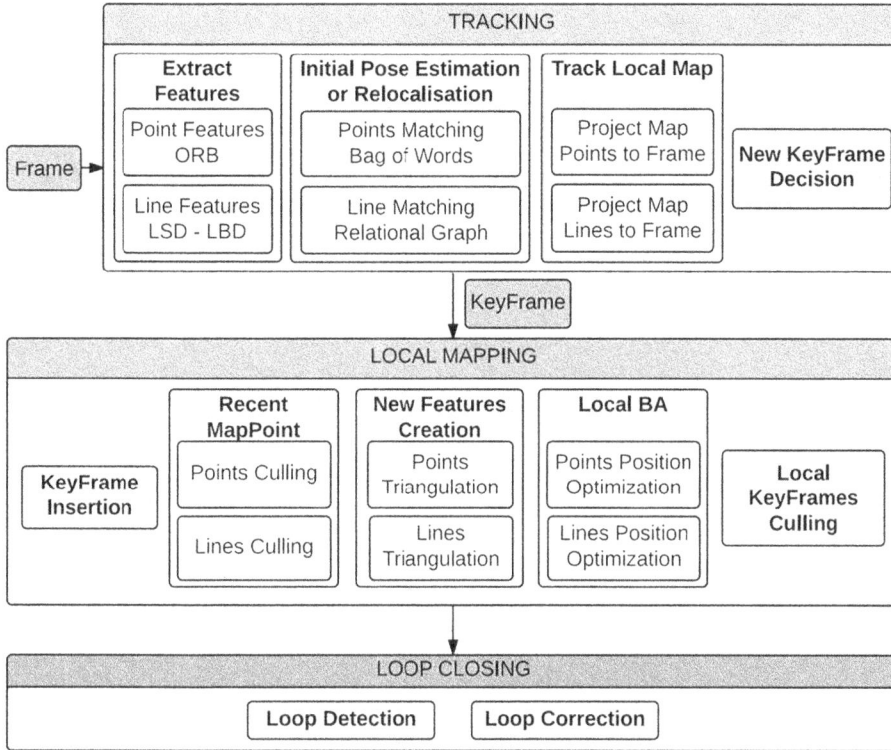

FIGURE 1.6   The framework of PL-SLAM [10].

leverages most of the ORB-SLAM [7] architecture and extends ORB-SLAM by integrating line features. The system consists of three threads: tracking, local mapping, and loop closing. In the tracking thread, on the basis of original ORB point features, line features are detected by means of LSD [21], and their local appearances are described by line band descriptors (LBD). The extracted line features are further matched with lines already present in the map using a relational graph strategy [22], which is combined with the bag-of-words-based points matching for the initial pose estimation or relocalization. As it is done with point features, after obtaining an initial set of map-to-image line feature pairs, all lines of the local map are projected onto the image to find further correspondences. Then, if the image contains sufficient new information, it is flagged as a keyframe and delivered to the local mapping thread. The local mapping thread culls redundant points and lines according to the observation information, and then creates new point and line features by triangularizing the matched point and line pairs, respectively. Besides the point-based reprojection error in ORB-SLAM, a new line-based reprojection error is added to optimize the camera poses, point positions, and line positions in the local BA module. Since matching lines across the whole map is computationally expensive, only point features are used for loop detection in the loop closing thread. As a result, PL-SLAM maintains the real-time performance of ORB-SLAM while enhancing the overall SLAM accuracy and robustness.

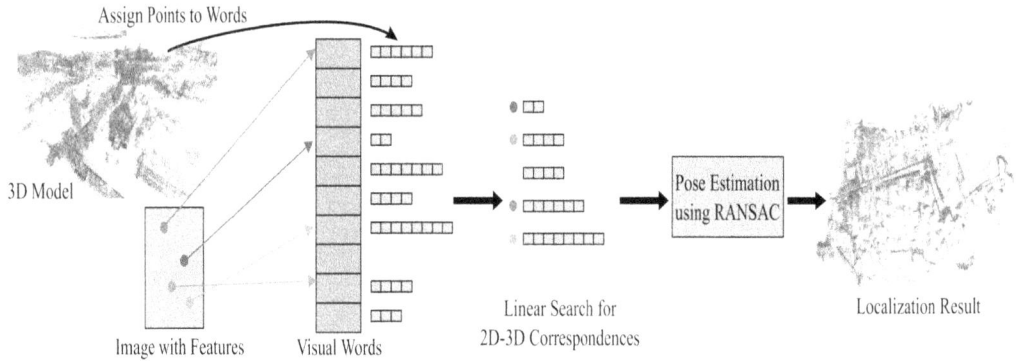

FIGURE 1.7   Illustration of VPS method [11].

1.2.5 VPS

In explicit map-based relocalization methods, 2D–3D feature matching is a crucial step. Sattler et al. [11] introduced a vocabulary-based prioritized search method (VPS) including visual vocabulary quantization and prioritized correspondence search to match the 2D image with the explicit map for relocalization. The framework of the VPS method is illustrated in Figure 1.7. The 3D model of scene is previously created by structure from motion [23], where the 3D model contains 3D points with their corresponding scale-invariant feature transform (SIFT) descriptors [24]. Note that a common quantized representation for the SIFT descriptors is utilized to convert floating point descriptor entries to integer values in the range [0, 255] for less memory. For faster indexing, the 3D points with their quantized SIFT descriptors are first stored in a quantized visual vocabulary obtained from an image set unrelated to the used datasets. During online localization, each feature in the query image is assigned to a visual word. A linear search is then conducted among all 3D points associated with that word to find two 3D points that best match the 2D feature point in the query image. The 2D–3D correspondence that passes the ratio test is accepted. To accelerate the 2D–3D matching process, a prioritized search strategy is proposed to process query features in ascending order of matching cost, where the cost is determined by the number of descriptors stored in the visual word of a query feature. Eventually, the matched 2D–3D point pairs are then used to solve for the camera pose using a RANSAC-based PnP method [15].

1.2.6 Hierarchical_loc

In large-scale environments, the phenomenon of perceptual aliasing often occurs where similar structures may appear at different locations, leading to many similar local features and thus making 2D–3D matching more difficult. Global image features, which capture a larger receptive field, are easier to distinguish between different locations compared to local features. Current explicit map-based visual localization methods for large-scale environments often begin with visual place recognition [25,26], where global image features are used to find candidate reference images in the map with feature distances close to the query image. 2D–3D matching is then performed according to the map points corresponding to

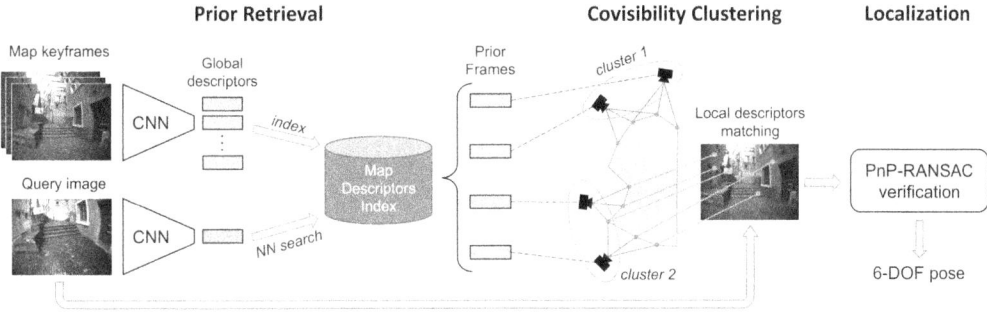

FIGURE 1.8   Overview of hierarchical_loc method [12].

these reference images to estimate the camera pose for the query image. The introduction of visual place recognition increases the scalability of localization, and it has become a research focus. Sarlin et al. [12] proposed an efficient hierarchical_loc framework combining visual place recognition and local 2D–3D matching. They first localize at the map level by place recognition with learned image-wide global descriptors, and subsequently estimate a precise pose from 2D–3D matches computed in the candidate places only. The detailed pipeline is depicted in Figure 1.8, which consists of three components: prior retrieval, covisibility clustering, and localization. The prior retrieval is a place recognition process, where a CNN network called MobileNetVLAD is trained by knowledge distillation [27] to extract the global descriptors for input images. Utilizing this trained CNN network, the global descriptors of map keyframes are first estimated and indexed to construct a map descriptors index. For each given query image, its corresponding descriptor is computed via the same CNN network and then a coarse nearest neighbors search is executed to find candidate prior frames in the map with most similar global descriptors. Next, these prior frames are clustered into places based on the 3D structure covisibility. These clusters are then iterated from the one containing the most prior frames, and local descriptor matching is executed between the 3D points in current cluster and 2D keypoints in the query image. The resulting 2D–3D correspondences are subsequently sent to the PnP-RANSAC verification module to compute a 6D pose in a RANSAC outlier rejection scheme. If a valid pose is found, the process terminates and the query image is successfully localized. Otherwise, it iterates to the next cluster, until a valid pose is found or until all clusters have been examined. Overall, the number of 3D points for matching is significantly smaller than the total number of points in the map, enabling the use of high-dimensional descriptors for robust matching with a tractable runtime.

1.2.7 HSCNet

Li et al. [13] proposed a hierarchical scene coordinate classification and regression network (HSCNet) to predict pixel scene coordinates in a coarse-to-fine manner from a single RGB image. As shown in Figure 1.9, this network consists of a series of output layers, each of which is conditioned on the previous ones. The final output layer predicts the 3D coordinates and the others produce progressively finer discrete location prediction.

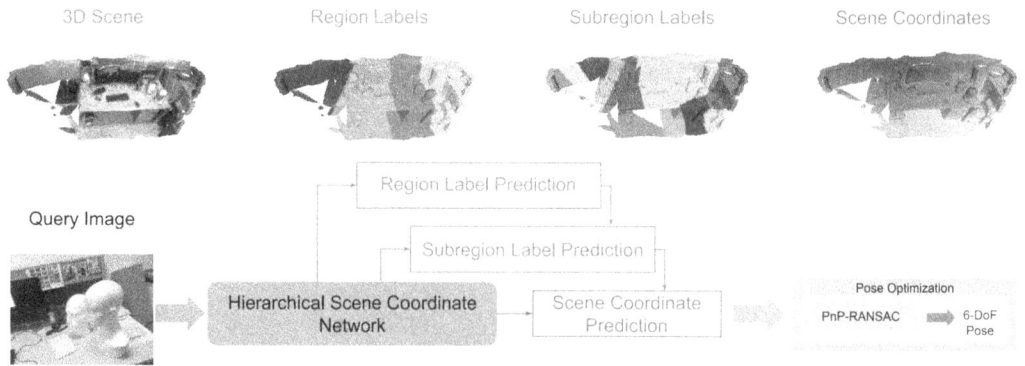

FIGURE 1.9   The system overview of HSCNet [13].

The dense 3D scene model is provided as the ground-truth 3D scene coordinates, and it is partitioned hierarchically with k-means to obtain the hierarchical discrete location labels. Take three output layers as an example, the discrete location labels contain region and subregion labels. Consequently, each pixel in a training image is associated with three-level labels: region label, subregion label, and 3D scene coordinate. For each of the front two levels, the proposed HSCNet has a corresponding classification layer to predict the pixel-wise locations at that level. In addition, a regression layer is designed inspired by [28] for the third level to predict the continuous 3D scene coordinates for each pixel. To enhance the connection of different levels, HSCNet propagates the coarse location information of the previous layer by a conditioning layer to guide the finer location predictions at the next levels. Specifically, the coarse label from the previous layer is utilized to generate a set of conditioning parameters with scaling and shifting, which are sent to the conditioning layer to impose linear transformation to the input feature map at the next level. Finally, the continuous 3D scene coordinates output from the final layer is combined with their corresponding 2D pixel coordinates to perform a PnP-RANSAC for 6D pose estimation.

1.2.8 PoseNet

PoseNet [14] was one of the earliest works to apply deep convolutional neural networks to end-to-end 6D camera pose estimation. It extracts feature vectors from input images using a feature extractor based on a slightly modified version of GoogLeNet [29] with 23 layers and maps these feature vectors to the pose space through a fully connected network. The output of PoseNet is a pose vector including a 3D camera position and orientation represented by a quaternion, where the quaternion representation is selected to simplify the legitimation of rotation by normalizing the orientation result to unit length. For the network training, the training labels (ground-truth camera poses) are automatically generated from scene videos using the structure from motion technique. The network parameters pre-trained on a large-scale image classification dataset are transferred to the pose regression task. In this process, the pre-trained parameters are used to initialize the feature extractor, while task-specific layers, such as fully connected layers for pose regression, are

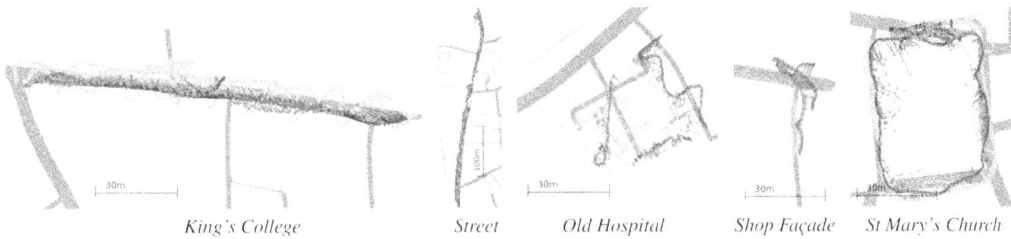

King's College   Street   Old Hospital   Shop Façade   St Mary's Church

FIGURE 1.10   Map of dataset [14].

newly initialized. The entire network is then fine-tuned using a pose regression dataset to optimize for the 6D camera pose prediction task.

Such an approach reduces the dependence on the amount of training data to some extent. Figure 1.10 displays a bird's eye view of the camera poses. Although the train and test images are taken from distinct walking paths, the predicted camera poses are still close to the ground-truth pose of test frames.

1.3 REVIEW OF LiDAR SLAM/ODOMETRY

LiDAR SLAM/odometry calculates the relative motion of the LiDAR sensor and the pose of source frame is estimated by aligning the point cloud of source frame to target data, where the target data can be from the last frame [30] or aggregated from multiple historical frames [31]. There are two main types: feature-based and distribution-based one. The former extracts features such as point, line, plane features from source frame for source-target matching and pose estimation. The latter represents the point cloud in the form of distribution, and the pose is estimated by maximizing the likelihood probability that is the product of observation probabilities from source-target correspondences. The feature-based methods focus on valuable environmental structures with good accuracy [32–35], while the distribution-based solution endeavors to improve the accuracy by involving more original LiDAR points at the expense of computation complexity [36–38].

1.3.1 LOAM

Zhang and Singh [32] proposed a milestone method termed as 3D LiDAR odometry and mapping (LOAM), which realizes real-time and precise localization by combining scan-to-scan matching module (LiDAR odometry) and scan-to-map matching module (LiDAR mapping), as shown in Figure 1.11a. This method concerns two type of geometric features: edge and planar point features, where the edge point is extracted from linear structures in the environment and the planar one is from regions with good planarity. Specifically, LOAM extracts the point features by evaluating the smoothness of the local surface for each point. A point with good smoothness is regarded as a planar point feature and an edge point feature is with weak smoothness. For scan-to-scan matching, the source frame is current frame and the target frame is the previous frame. For each point feature in the source frame, its nearest point corresponding to the source projection in the target frame is obtained. Further, the neighboring points of the nearest point in the adjacent scan lines are searched. These searched points are then fitted to a line or

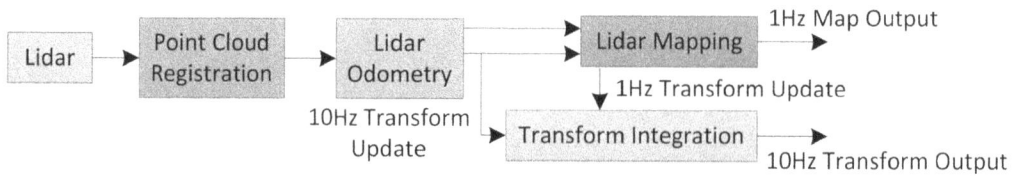

(a)

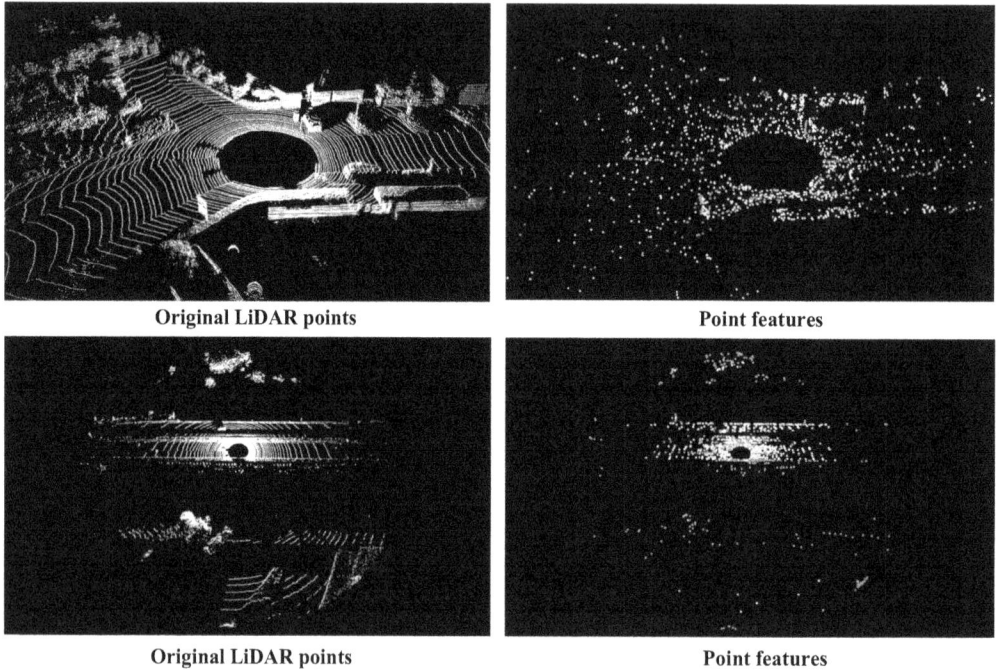

(b)

FIGURE 1.11 LOAM [32]. (a) The framework of LOAM. (b) The original point cloud and the extracted edge and planar points.

a plane corresponding to source edge or planar point feature, respectively. On this basis, the constraints of point-to-line and point-to-plane distances are formed. By minimizing the sum of these constraints, the pose of scan-to-scan matching is estimated using the Levenberg–Marquardt algorithm [39]. The scan-to-scan matching ensures computational efficiency at 10 Hz with a poor estimation accuracy, which is promoted by scan-to-map matching. The source point features of current frame are projected into the target feature map, where the nearest 5 point features of each projection are searched and fitted to a line or a plane. Similarly, the constraints of point-to-line and point-to-plane distances are constructed for pose optimization. It runs at the frequency of 1 Hz. To ensure real-time and good localization result, the estimated poses of these two modules are combined in the transform integration module, where the result from scan-to-scan matching is appropriately interpolated into that of scan-to-map matching. And the final pose is outputted at 10 Hz. Figure 1.11b provides examples of edge and planar point feature extraction for urban and highway scenes on the KITTI dataset. Considering that LOAM is not open-source,

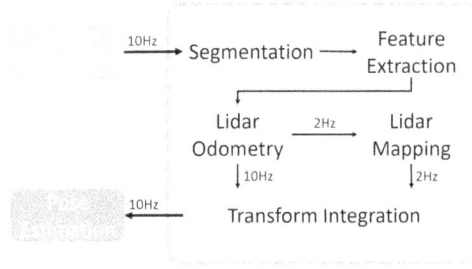

FIGURE 1.12   The system overview of LeGO-LOAM [33].

we choose A-LOAM [40] to provide the point features in Figure 1.11b, where A-LOAM is generally considered as an open-source implementation of LOAM, which uses eigen [41] and ceres solver [42] to simplify code structure. The results show the adaptability of the feature extraction to different scenes.

1.3.2 LeGO-LOAM

On the basis of LOAM, Shan and Englot [33] proposed a lightweight and ground-optimized version called LeGO-LOAM, as shown in Figure 1.12. There are mainly three significant improvements. Firstly, it incorporates a segmentation module prior to feature extraction. Thus, the point cloud is clustered, and ground and non-ground points can be classified. The non-ground cluster with less points is discarded for effectively removing noise such as points of tree leaves. Secondly, the six degrees of freedom of the pose are estimated by dividing them into two sets of three degrees of freedom $\left(t_z, \theta_{\mathrm{roll}}, \theta_{\mathrm{pitch}}\right)$ and $\left(t_x, t_y, \theta_{\mathrm{yaw}}\right)$, where t_x, t_y, and t_z are translational components, and $\theta_{\mathrm{roll}}, \theta_{\mathrm{pitch}}$, and θ_{yaw} refer to rotational components. The former is optimized only using planar point features, while the latter is optimized by utilizing both edge and planar point features. Such division is beneficial to reduce the optimization time. Thirdly, the loop closure detection and pose-graph optimization are introduced, which enhances the global consistency of trajectory. It is noted that the processing frequency of scan-to-map matching is lifted to 2 Hz, which is integrated with 10 Hz scan-to-scan matching similar to LOAM. Still, the pose output frequency of LeGO-LOAM is 10 Hz.

1.3.3 SuMa

Behley and Stachniss [34] proposed an efficient surfel-based mapping (SuMa) method, as shown in Figure 1.13. Each surfel (surface element) is a feature to describe a local surface, which is composed of a position, a normal vector, and a radius. Also, it carries two timestamps: the creation timestamp and the last update timestamp. After preprocessing of point cloud, a vertex map $\mathcal{V}_D$ and a normal map $\mathcal{N}_D$ are generated. Afterward, a projective data association is applied by projecting target surfels to the vertex map, and the surfel correspondences between the current frame and the active surfel map $\mathcal{M}_{\mathrm{active}}$ are found. On this basis, the constraints of point-to-plane distances are used to form the pose optimization function to estimate the pose T_{WC_t}. Furthermore, the loop detection and pose-graph optimization are performed in a separate thread with the inactive surfel map $\mathcal{M}_{\mathrm{inactive}}$.

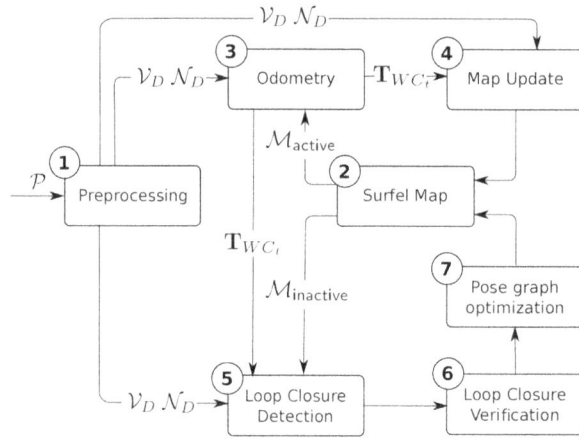

FIGURE 1.13   The framework of SuMa [34].

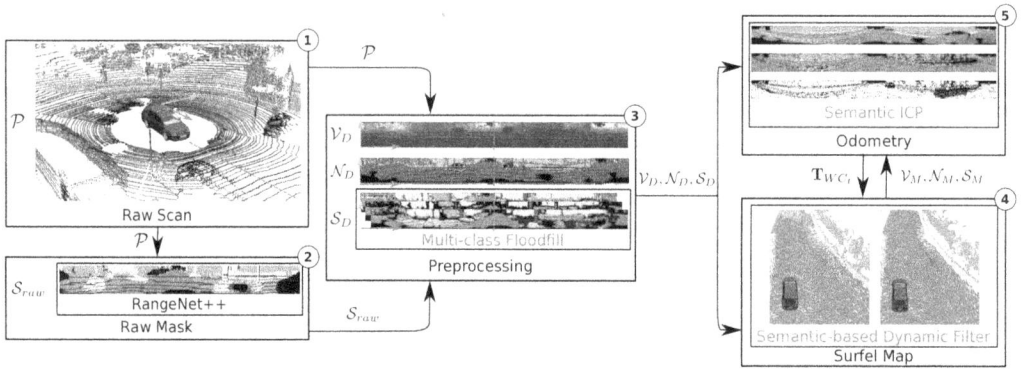

FIGURE 1.14   The framework of SuMa++ [35].

1.3.4 SuMa++

SuMa++ [35] is an extension of SuMa, which integrates semantic information to enhance the mapping quality and improve the accuracy of localization in dynamic environments. The framework of SuMa++ is shown in Figure 1.14. For a raw scan, it is transformed into range image by the spherical projection and the point-wise semantic information is extracted by a convolutional neural network RangeNet++. Similar to SuMa, the vertex map $\mathcal{V}_D$ and a normal map $\mathcal{N}_D$ are generated. Besides, the semantic map $\mathcal{S}_D$ is also produced in the preprocessing module using multi-class flood-fill. The flood-fill algorithm is utilized to reduce errors of the semantic labels when the labels are re-projected to the map, due to the projective input and the blob-like outputs produced as a by-product of in-network down-sampling of RangeNet++. With the semantic surfels of current frame and the semantic surfel map, the semantic iterative closest point (ICP) is performed to estimate the pose T_{WC_t}, where the constraint of point-to-plane distance is weighted by the semantic consistency between the associated surfels. This makes pose estimation more robust to outliers.

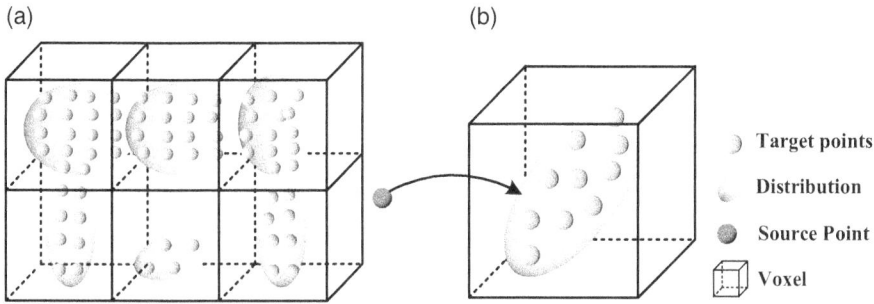

FIGURE 1.15 The distribution representation and association of NDT. (a) The distribution representation. (b) The point-to-voxel association.

During the map updating process, a semantic-based dynamic filter is added, which checks the semantic consistency between the current frame and the semantic surfel map. Such a filter alleviates the interference caused by dynamic objects in the actual environment.

1.3.5 Normal Distribution Transform

The normal distribution transform (NDT) [36] is a robust method for registering source point cloud to the target map, where the target map is discretized as a voxel map, as shown in Figure 1.15a. Each voxel is modeled as a three-dimensional Gaussian distribution with the mean and covariance matrix formed by points inside the voxel. Then, the source point cloud is projected into the voxel map, and each source point and the voxel where the source point projection is located form a point-to-voxel association, as shown in Figure 1.15b. The core of the NDT is an iterative optimization process, which optimizes the pose of source frame to minimize the mismatch between the Gaussian distributions of target map and the source points. This mismatch is quantified using the Mahalanobis distance that measures how well each source point fits the Gaussian distribution of its associated voxel. Finally, the optimized pose is obtained.

1.3.6 LiTAMIN

LiTAMIN (LiDAR-based tracking and mapping) [37] is another distribution-based method. It improves accuracy, robustness, and computational efficiency of the ICP algorithm by employing a locally approximated geometry with clusters of normal distributions. The LiDAR scan is input into the odometry module, which contains two threads: a pose tracking thread and a local mapping thread. This module computes the pose of current frame by registering with the local map, where the local map is built with historical scans by the tracking thread. On this basis, the keyframe maker accumulates the poses and the keyframe of every 10 m is determined, where the memory usage is reduced by writing the keyframe to storage, such as a hard drive or solid-state drive. Then, the pose-graph optimizer detects loops in the pose graph and corrects the recent relative poses among keyframes with the ICP-based loop closure processing. Through pose-graph optimization, robust and accurate localization is achieved (Figure 1.16).

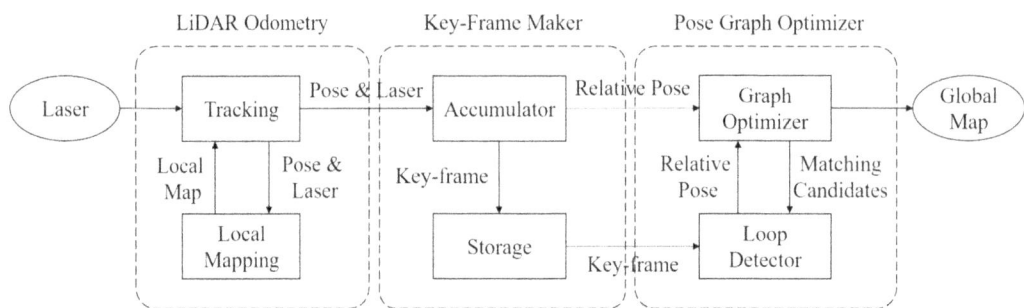

FIGURE 1.16 The system overview of LiTAMIN [37].

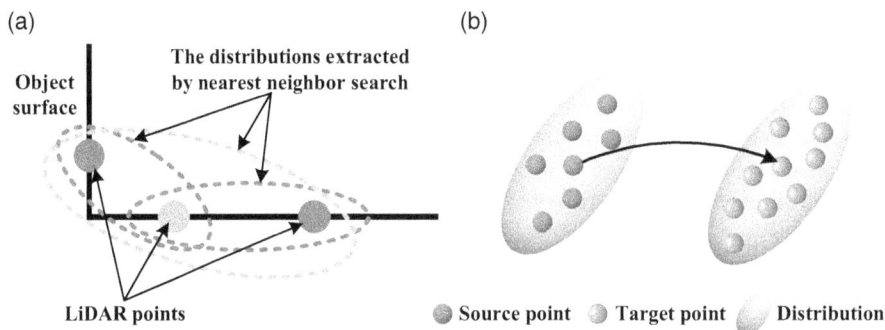

FIGURE 1.17 The distribution representation and association of GICP. (a) The distribution representation. (b) The point-to-point association.

1.3.7 Generalized Iterative Closest Point (GICP)

Generalized iterative closest point (GICP) [38] is an advanced point cloud registration algorithm where the source and target point cloud are both modeled as a series of Gaussian distributions. Specifically, for a point in a frame, its neighboring points are searched by nearest neighbor search to calculate the Gaussian distribution with the mean and covariance matrix, as illustrated in Figure 1.17a. After the distributions are extracted, the source distributions are projected to the target frame, and the point-to-point association is performed between source and target distributions, as shown in Figure 1.17b. The distance difference between the source projection point and the matched target point is assumed to obey the difference between the corresponding Gaussian distributions. This determines the cost function for pose optimization, which is solved in an iterative way until convergence.

1.4 OVERVIEW OF THE SUBSEQUENT CHAPTERS

In Chapter 2, the mathematical background and basic localization and mapping theory are provided. Chapter 3 explores the semantic visual SLAM approach with point and object features when the environment map is unknown. In the case that environment map is known, Chapter 4 focuses on visual relocalization from the perspective of scene coordinate regression network for accurate localization of small scenes, while Chapter 5 concerns on visual relocalization from the perspective of visual place recognition for scalability to large-scale scenes. On this basis, Chapter 6 proposes a localization software architecture for service robots based on the hybrid map, which integrates the above-presented three localization methods together under the ROS framework. The reliable and stable localization is then achieved, laying a foundation for the robot task execution. In the following section, the LiDAR localization is addressed. Chapter 7 presents the research on LiDAR odometry and Chapter 8 further combines IMU to form LiDAR inertial odometry for robust localization. LiDAR place recognition is detailed in Chapters 9, and Chapter 10 gives a summary and outlook.

REFERENCES

[1] Kim, M., Kang, T., Song, D., & Yi, S. J. (2021). Development of a small-sized intelligent home service robot. In *Proceedings of the International Conference on Ubiquitous Robots, Gangneung*, Korea (South) (pp. 565–570).

[2] Wang, Z., Liao, H., Jia, Z., & Wu, J. (2022). Semantic mapping based on visual SLAM with object model replacement visualization for cleaning robot. In *Proceedings of the IEEE International Conference on Robotics and Biomimetics*, Jinghong, China (pp. 569–575).

[3] Putra, P. D. H., Riansyah, M. I., & Farouq, A. A. (2022). Localization design and implementation using combined rotary encoder, IMU, and ROS on delivery service robots in building area. In *Proceedings of the 8th International Conference on Science and Technology*, Yogyakarta, Indonesia, doi: 10.1109/ICST56971.2022.10136264.

[4] Shi, Z., Chang, X., Yang, C., Wu, Z., & Wu, J. (2020). An acoustic-based surveillance system for amateur drones detection and localization. *IEEE Transactions on Vehicular Technology*, 69(3), 2731–2739.

[5] Engel, J., Schöps, T., & Cremers, D. (2014). LSD-SLAM: Large-scale direct monocular SLAM. In *Proceedings of European Conference on Computer Vision*, Zurich, Switzerland (pp. 834–849).

[6] Engel, J., Koltun, V., & Cremers, D. (2017). Direct sparse odometry. *IEEE Transactions on Pattern Analysis and Machine Intelligence*, 40(3), 611–625.

[7] Mur-Artal, R., Montiel, J. M. M., & Tardos, J. D. (2015). ORB-SLAM: A versatile and accurate monocular SLAM system. *IEEE Transactions on Robotics*, 31(5), 1147–1163.

[8] Mur-Artal, R., & Tardós, J. D. (2017). ORB-SLAM2: An open-source SLAM system for monocular, stereo, and RGB-D cameras. *IEEE Transactions on Robotics*, 33(5), 1255–1262.

[9] Campos, C., Elvira, R., Rodríguez, J. J. G., Montiel, J. M., & Tardós, J. D. (2021). ORB-SLAM3: An accurate open-source library for visual, visual-inertial, and multimap SLAM. *IEEE Transactions on Robotics*, 37(6), 1874–1890.

[10] Pumarola, A., Vakhitov, A., Agudo, A., Sanfeliu, A., & Moreno-Noguer, F. (2017). PL-SLAM: Real-time monocular visual SLAM with points and lines. In *Proceedings of IEEE International Conference on Robotics and Automation*, Singapore (pp. 4503–4508).

[11] Sattler, T., Leibe, B., & Kobbelt, L. (2011). Fast image-based localization using direct 2D-to-3D matching. In *Proceedings of the International Conference on Computer Vision*, Barcelona, Spain (pp. 667–674).

[12] Sarlin, P. E., Debraine, F., Dymczyk, M., Siegwart, R., & Cadena, C. (2018). Leveraging deep visual descriptors for hierarchical efficient localization. In *Proceedings of the Conference on Robot Learning*, Zurich, Switzerland (pp. 456–465).

[13] Li, X., Wang, S., Zhao, Y., Verbeek, J., & Kannala, J. (2020). Hierarchical scene coordinate classification and regression for visual localization. In *Proceedings of the IEEE Conference on Computer Vision and Pattern Recognition*, Seattle, WA, USA (pp. 11983–11992).

[14] Kendall, A., Grimes, M., & Cipolla, R. (2015). PoseNet: A convolutional network for real-time 6-DOF camera relocalization. In *Proceedings of the IEEE International Conference on Computer Vision*, Santiago, Chile (pp. 2938–2946).

[15] Hartley, R., & Zisserman, A. (2003). *Multiple View Geometry in Computer Vision*. Cambridge University Press.

[16] Engel, J., Sturm, J., & Cremers, D. (2013). Semi-dense visual odometry for a monocular camera. In *Proceedings of the IEEE International Conference on Computer Vision*, Sydney, NSW, Australia (pp. 1449–1456).

[17] Engel, J., Usenko, V., & Cremers, D. (2016). A photometrically calibrated benchmark for monocular visual odometry. arXiv preprint arXiv:1607.02555.

[18] Rublee, E., Rabaud, V., Konolige, K., & Bradski, G. (2011). ORB: An efficient alternative to SIFT or SURF. In *Proceedings of International Conference on Computer Vision*, Barcelona, Spain (pp. 2564–2571).

[19] Gálvez-López, D., & Tardos, J. D. (2012). Bags of binary words for fast place recognition in image sequences. *IEEE Transactions on Robotics*, 28(5), 1188–1197.

[20] Forster, C., Carlone, L., Dellaert, F., & Scaramuzza, D. (2016). On-manifold preintegration for real-time visual-inertial odometry. *IEEE Transactions on Robotics*, 33(1), 1–21.

[21] Von Gioi, R. G., Jakubowicz, J., Morel, J. M., & Randall, G. (2012). LSD: A line segment detector. *Image Processing On Line*, 2, 35–55.

[22] Zhang, L., & Koch, R. (2013). An efficient and robust line segment matching approach based on LBD descriptor and pairwise geometric consistency. *Journal of Visual Communication and Image Representation*, 24(7), 794–805.

[23] Snavely, N., Seitz, S. M., & Szeliski, R. (2008). Modeling the world from internet photo collections. *International Journal of Computer Vision*, 80(2), 189–210.

[24] Lowe, D. G. (2004). Distinctive image features from scale-invariant keypoints. *International Journal of Computer Vision*, 60(2), 91–110.

[25] Lowry, S., Sünderhauf, N., Newman, P., Leonard, J. J., Cox, D., Corke, P., & Milford, M. J. (2015). Visual place recognition: A survey. *IEEE Transactions on Robotics*, 32(1), 1–19.

[26] Zhang, X., Wang, L., & Su, Y. (2021). Visual place recognition: A survey from deep learning perspective. *Pattern Recognition*, 113, 107760.

[27] Hinton, G. (2015). Distilling the knowledge in a neural network. arXiv preprint arXiv: 1503.02531.

[28] Perez, E., Strub, F., De Vries, H., Dumoulin, V., & Courville, A. (2018). Film: Visual reasoning with a general conditioning layer. In *Proceedings of the AAAI Conference on Artificial Intelligence*, 32(1), 3942–3951.

[29] Szegedy, C., Liu, W., Jia, Y., Sermanet, P., Reed, S., Anguelov, D., Erhan, D., Vanhoucke, V., & Rabinovich, A. (2015). Going deeper with convolutions. In *Proceedings of the IEEE Conference on Computer Vision and Pattern Recognition*, Boston, MA, USA (pp. 1–9), doi: 10.1109/CVPR.2015.7298594.

[30] Koide, K., Yokozuka, M., Oishi, S., & Banno, A. (2021). Voxelized GICP for fast and accurate 3D point cloud registration. In *Proceedings of IEEE International Conference on Robotics and Automation*, Xi'an, China (pp. 11054–11059).

[31] Chang, D., Zhang, R., Huang, S., Hu, M., Ding, R., & Qin, X. (2023). WiCRF: Weighted bimodal constrained LiDAR odometry and mapping with robust features. *IEEE Robotics and Automation Letters*, 8(3), 1423–1430.

[32] Zhang, J., & Singh, S. (2014). LOAM: Lidar odometry and mapping in real-time. In *Robotics: Science and Systems*, doi: 10.15607/RSS.2014.X.007.

[33] Shan, T., & Englot, B. (2018). LeGO-LOAM: Lightweight and ground-optimized lidar odometry and mapping on variable terrain. In *Proceedings of the IEEE/RSJ International Conference on Intelligent Robots and Systems*, Madrid, Spain (pp. 4758–4765).

[34] Behley, J., & Stachniss, C. (2018). Efficient surfel-based SLAM using 3D laser range data in urban environments. In *Robotics: Science and Systems*, doi: 10.15607/RSS.2018.XIV.016.

[35] Chen, X., Milioto, A., Palazzolo, E., Giguère, P., Behley, J., & Stachniss, C. (2019). SuMa++: Efficient LiDAR-based semantic SLAM. In *Proceedings of the IEEE/RSJ International Conference on Intelligent Robots and Systems*, Macau, China (pp. 4530–4537).

[36] Magnusson, M., Lilienthal, A., & Tom, D. (2007). Scan registration for autonomous mining vehicles using 3D-NDT. *Journal of Field Robotics*, 24, 803–827.

[37] Yokozuka, M., Koide, K., Oishi, S., & Banno, A. (2020). LiTAMIN: LiDAR-based tracking and mapping by stabilized ICP for geometry approximation with normal distributions. In *Proceedings of the IEEE/RSJ International Conference on Intelligent Robots and Systems*, Las Vegas, NV, USA (pp. 5143–5150).

[38] Segal, A., Haehnel, D., & Thrun, S. (2009). Generalized-ICP. In *Robotics: Science and Systems*, doi: 10.15607/RSS.2009.V.021.

[39] Grisetti, G., Kümmerle, R., Stachniss, C., & Burgard, W. (2010). A tutorial on graph-based SLAM. *IEEE Intelligent Transportation Systems Magazine*, 2(4), 31–43.

[40] Qin, T., & Cao, S. https://github.com/HKUST-Aerial-Robotics/A-LOAM.

[41] Guennebaud, G., Jacob, B., et al. https://github.com/PX4/eigen.

[42] Agarwal, S., Mierle, K., et al. https://github.com/ceres-solver.

CHAPTER 2

Mathematical Foundation of Localization and Mapping Theory

2.1 INTRODUCTION

Localization and mapping are the cornerstone of autonomous movement and navigation for mobile robots, and their success relies on a solid mathematical foundation. This chapter provides a comprehensive overview of the key mathematical tools and theoretical underpinnings that support localization and mapping algorithms, which bridges the gap between abstract mathematical concepts and their practical application in solving real-world localization and mapping problems.

The key to localization and mapping is motion estimation of robots. We begin with the representation of 3D rigid body motion in Section 2.1, focusing on different representation forms and their conversions. In practice, the pose transformation after a motion is obtained based on the sensor observations, where the sensor model needs to be established to associate the 3D objects and their corresponding observation. Hence, Section 2.2 gives the common pinhole camera model and LiDAR model. According to the correspondence relationship of sensor observations, Section 2.3 presents motion estimation approaches. Finally, Section 2.4 concludes this chapter.

2.2 3D RIGID BODY MOTION

The motion of a rigid body in 3D space consists of two fundamental components: rotation and translation. Translation is actually the distance the rigid body moves along three axes, which is relatively straightforward. However, handling rotations poses a more complex challenge. Two widely used representations for rotations are rotation matrices [1] and rotation vectors [2], each offering a unique advantage and suited for different computational needs. The rotation matrix offers a complete and straightforward way to describe rotation, while the rotation vector provides a more compact alternative. Next, we will

DOI: 10.1201/9781003643630-2

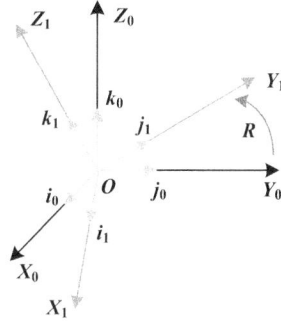

FIGURE 2.1   Illustration of rotation transformation.

introduce the definitions of rotation matrix, rotation vector, and how they are calculated and transformed.

2.2.1 Rotation Matrix

The rotation component of robot motion is considered in Figure 2.1. Assume that a robot undergoes rotation motion, the robot coordinate system is rotated from $O_1 X_1 Y_1 Z_1$ to $O_2 X_2 Y_2 Z_2$. The corresponding orthogonal unit bases of $O_1 X_1 Y_1 Z_1$ and $O_2 X_2 Y_2 Z_2$ are denoted as (i_0, j_0, k_0) and (i_1, j_1, k_1), respectively. For a space vector p, we assume that its coordinates in the two coordinate system are $[x_0, y_0, z_0]^T$ and $[x_1, y_1, z_1]^T$. Since vector p is fixed, we can obtain that [3]

$$p = \begin{bmatrix} i_0, j_0, k_0 \end{bmatrix} \begin{bmatrix} x_0 \\ y_0 \\ z_0 \end{bmatrix} = \begin{bmatrix} i_1, j_1, k_1 \end{bmatrix} \begin{bmatrix} x_1 \\ y_1 \\ z_1 \end{bmatrix} \tag{2.1}$$

Left multiply both side with $\begin{bmatrix} i_0^T \\ j_0^T \\ k_0^T \end{bmatrix}$ and consider the orthogonality between the basis vectors, we have

$$\begin{bmatrix} x_0 \\ y_0 \\ z_0 \end{bmatrix} = \begin{bmatrix} i_0^T i_1 & i_0^T j_1 & i_0^T k_1 \\ j_0^T i_1 & j_0^T j_1 & j_0^T k_1 \\ k_0^T i_1 & k_0^T j_1 & k_0^T k_1 \end{bmatrix} \begin{bmatrix} x_1 \\ y_1 \\ z_1 \end{bmatrix} \tag{2.2}$$

Define matrix $R = \begin{bmatrix} i_0^T i_1 & i_0^T j_1 & i_0^T k_1 \\ j_0^T i_1 & j_0^T j_1 & j_0^T k_1 \\ k_0^T i_1 & k_0^T j_1 & k_0^T k_1 \end{bmatrix}$, which describes the rotation motion of the robot and thus is called as rotation matrix. Meantime, R also characterizes the coordinate transformation of the same vector in different coordinate systems. Note that both $\left(i_0, j_0, k_0\right)$ and $\left(i_1, j_1, k_1\right)$ are orthonormal bases, so each element of rotation matrix R is the cosine of the angle between the base vectors of two coordinate systems.

The properties of the 3D rotation matrix are as follows [3]:

1. **Orthogonality**: A 3D rotation matrix R is an orthogonal matrix, meaning that its rows and columns are orthonormal vectors. Mathematically, this is expressed as: $R^T R = RR^T = I$, where I is the identity matrix.

2. **Determinant**: The determinant of a 3D rotation matrix is always 1: $\det(R) = 1$.

3. **Preservation of length and angle**: A rotation matrix preserves the length of vectors and the angles between them. If a vector a is rotated by the matrix R, the length of the vector remains the same: $|Ra| = |a|$. Similarly, the angle between any two vectors is preserved under rotation.

4. **Composition of rotations**: The product of two rotation matrices is also a rotation matrix. This means that consecutive rotations can be represented by the multiplication of their respective rotation matrices: $R_1 R_2$ represents the result of applying the rotation R_2 followed by R_1.

2.2.2 Rotation Vector

In the above section, we use rotation matrix with nine elements to represent a 3D rotation motion with 3 degrees of freedom, and thus, such an expression is redundant. In fact, a rotation motion can be described using a rotation axis (3D unit vector) u and a rotation angle (scalar) θ. Therefore, we can represent 3D rotation motion using a 3D vector θu, whose direction is consistent with that of rotation axis and length represents the rotation angle [2]. Such representation of describing the rotation using an axis and angle is called the rotation vector representation (or axis-angle representation).

Next, we will derive how the rotation vector acts on a vector in 3D space. Given a vector v in 3D space, we want to rotate it by an angle θ around the unit vector u. First, the vector v can be decomposed into two components: one perpendicular to u, denoted as $v_\perp$, and one parallel to u, denoted as $v_\parallel$, as shown in Figure 2.2a. Note that Figure 2.2b illustrates the

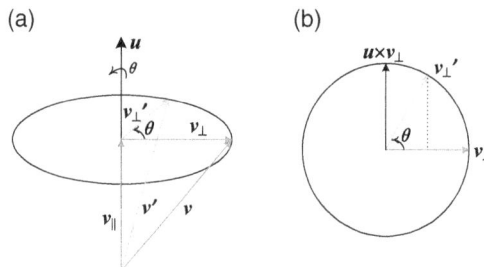

FIGURE 2.2 Illustration of rotation transformation with rotation vector. (a) Rotation transformation on vector v in the 3D space. (b) Top view of (a).

top view of Figure 2.2a. Based on the geometric relationships shown in Figure 2.2a, we can derive the following formulas [4]:

$$v_\parallel = (u \cdot v)u$$
$$v_\perp = v - v_\parallel = v - (u \cdot v)u$$

(2.3)

After the rotation, the component $v_\parallel$ does not change, while the perpendicular component $v_\perp$ undergoes rotation and becomes a new component $v'_\perp$. It can be seen from Figure 2.2 that

$$v'_\perp = \cos\theta v_\perp + \sin\theta(u \times v_\perp)$$

(2.4)

Combining (2.3) and (2.4), we can conclude that

$$v'_\perp = \cos\theta v_\perp + \sin\theta(u \times v)$$

(2.5)

The rotated vector v' is the sum of the rotated parallel component and rotated perpendicular component:

$$
\begin{aligned}
v' &= v_\parallel + v'_\perp \\
&= v_\parallel + \cos\theta v_\perp + \sin\theta(u \times v) \\
&= (u \cdot v)u + \cos\theta(v - (u \cdot v)u) + \sin\theta(u \times v) \\
&= \cos\theta v + (1 - \cos\theta)(u \cdot v)u + \sin\theta(u \times v)
\end{aligned}
$$

(2.6)

As a result, Equation (2.6) represents the result of rotating a spatial vector v using rotation vector θu.

2.2.3 Conversion between Rotation Vector and Rotation Matrix

Based on the derivation of the rotation vector in (2.6), we can further simplify it to obtain the following result:

$$
\begin{aligned}
v' &= \cos\theta v + (1 - \cos\theta)(u \cdot v)u + \sin\theta(u \times v) \\
&= \cos\theta v + (1 - \cos\theta)uu^T v + \sin\theta(u \times v) \\
&= \left[\cos\theta I + (1 - \cos\theta)uu^T + \sin\theta(u^\wedge)\right]v
\end{aligned}
$$

(2.7)

Comparing (2.7) with (2.2), one can see that

$$R = \cos\theta I + (1 - \cos\theta)uu^T + \sin\theta(u^\wedge)$$

(2.8)

where the symbol $u^{\wedge}$ represents the skew-symmetric matrix of vector u. Thus, we can convert the rotation vector into a rotation matrix using (2.8), which is known as Rodrigues' formula.

Alternatively, the rotation matrix can be converted into a rotation vector by applying the trace operation to both sides of Equation (2.8):

$$\operatorname{tr}(R) = \cos\theta\ \operatorname{tr}(I) + (1 - \cos\theta)\operatorname{tr}(uu^T) + \sin\theta\operatorname{tr}(u^{\wedge}) = 1 + 2\cos\theta \qquad (2.9)$$

Therefore, the angle of the rotation vector can be obtained as

$$\theta = \arccos\left(\frac{\operatorname{tr}(R) - 1}{2}\right) \qquad (2.10)$$

Since the result of vector u rotating around the rotation axis u remains unchanged, we can conclude that $Ru = u$. By solving this equation and normalization, we can obtain the rotation axis u. Combining (2.10), the final rotation vector is attained from rotation matrix.

2.3 SENSOR MODEL

2.3.1 Pinhole Camera Model

The pinhole camera model [5] is a simplified optical imaging model, where the basic principle is central perspective projection, projecting objects from 3D space onto a 2D imaging plane through a pinhole. Next, we take the pinhole camera model as an example to explore the mapping relationship between points in 3D space and the pixel points on an image.

Given a 3D point $P_c = (X_c, Y_c, Z_c)$ in the camera coordinate system C, it is first projected onto the corresponding imaging plane of the camera O_c (see Figure 2.3), and then the coordinate (x, y) of projected point P_i is further transformed into 2D pixel coordinate $p = (u, v)$ on the image. Based on the similarity of triangles, we have

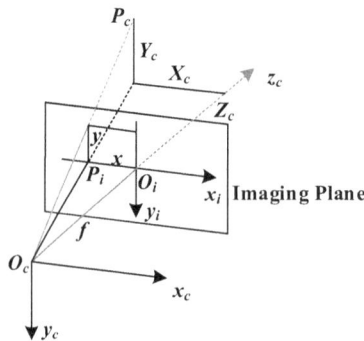

FIGURE 2.3 The projection transformation from 3D space in camera coordinate system to image plane. $O_c x_c y_c z_c$ and $O_i x_i y_i$ represent camera and image coordinate system, respectively. O_c and f correspond to the position of the camera's optical center and the focal length.

$$\frac{x}{X_c} = \frac{y}{Y_c} = \frac{f}{Z_c} \tag{2.11}$$

Therefore, we can conclude that

$$x = f\frac{X_c}{Z_c} \qquad y = f\frac{Y_c}{Z_c} \tag{2.12}$$

Equation (2.12) describes the spatial relationship between 3D point P_c and its projected point P_i on the imaging plane, where the coordinates of the two points are both in meters. However, a digital image is stored as a pixel matrix after discretization, where the coordinates are represented in pixels. To facilitate the conversion between spatial coordinates P_i and pixel coordinates p, we introduce a pixel coordinate system ouv, whose origin is located in the upper left corner of the image and u, v axes are respectively paralleled to x_i, y_i axes in Figure 2.3. By applying scaling and translation transformations on P_i, the pixel coordinate p in ouv system can be obtained as

$$\begin{cases} u = \alpha x + c_x \\ v = \beta y + c_y \end{cases} \tag{2.13}$$

Combining Equation (2.12), we have

$$\begin{cases} u = f_x \dfrac{X_c}{Z_c} + c_x \\ v = f_y \dfrac{Y_c}{Z_c} + c_y \end{cases} \tag{2.14}$$

where $f_x = \alpha f$ and $f_y = \beta f$. Write the above equation in matrix form:

$$Z_c \begin{bmatrix} u \\ v \\ 1 \end{bmatrix} = \begin{bmatrix} f_x & 0 & c_x \\ 0 & f_y & c_y \\ 0 & 0 & 1 \end{bmatrix} \begin{bmatrix} X_c \\ Y_c \\ Z_c \end{bmatrix} = KP_c \tag{2.15}$$

where $[u \ v \ 1]^T$ is the homogeneous form of pixel coordinate p, and $K = \begin{bmatrix} f_x & 0 & c_x \\ 0 & f_y & c_y \\ 0 & 0 & 1 \end{bmatrix}$ is the camera intrinsics [6].

In the camera coordinate system C, the origin is at the camera's optical center, and its axes align with the camera's orientation. As the camera moves, this coordinate system changes accordingly, which causes the same 3D point corresponding to various coordinates. However, in practice, a fixed reference frame called the world coordinate system W is often used. This stationary system provides a consistent reference for representing objects and scenes, enabling easier integration of multiple views from different camera positions. The transformation from the world coordinate system to camera coordinate system can be represented by a rotation matrix R_{cw} and a translation vector t_{cw}. Let P_w denote the corresponding 3D coordinate of P_c in the world coordinate system, and then Equation (2.15) can be written as

$$Z_c \begin{bmatrix} u \\ v \\ 1 \end{bmatrix} = K\left(R_{cw}P_w + t_{cw}\right) \tag{2.16}$$

where R_{cw} and t_{cw} are also referred to as the camera extrinsics. Compared with fixed intrinsics, camera extrinsics are changed with camera motion and denote the robot trajectory.

2.3.2 LiDAR Model

2.3.2.1 Measurement Model

The LiDAR sensor measures distance by continuously emitting laser beams, which reflect upon encountering obstacles. Some of these reflections are received again by the LiDAR sensor. By measuring the time taken for the beam to travel from the sensor to the obstacle and back (the round trip time (RTT) of the laser beam), it is possible to calculate the distance between the surrounding objects and the LiDAR sensor. The distance d from the LiDAR to the obstacle is determined using the formula:

$$d = \frac{Ct}{2} \tag{2.17}$$

where C is the speed of light and t is the RTT.

As shown in Figure 2.4, the measurement model of LiDAR is the Range-Azimuth-Elevation (RAE) model [7], where P is the observation point of the LiDAR sensor. d represents the distance from point P to the LiDAR sensor, which is computed from (2.17). α is the azimuth angle and ω is the elevation angle. Both α and ω are the emission angles

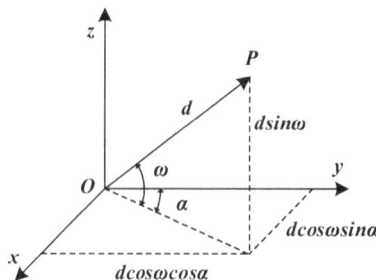

FIGURE 2.4   Illustration of LiDAR sensor model.

of the laser beam. Based on geometric relationships in Figure 2.4, the spatial coordinate of P can be expressed as

$$P = \begin{bmatrix} x \\ y \\ z \end{bmatrix} = \begin{bmatrix} d\cos\omega\sin\alpha \\ d\cos\omega\cos\alpha \\ d\sin\omega \end{bmatrix} \quad (2.18)$$

2.3.2.2 Distortion Correction

For most LiDAR systems, although the emission and reception of the laser occur very quickly, each point that constitutes the point cloud is not generated at the exact same moment. Generally, we consider the data accumulated within 100 ms as a frame of point cloud output. If within these 100 ms, the LiDAR unit itself or the body on which it is mounted undergoes a change in absolute position, then the coordinate system for each point in this frame of the point cloud will be different. Visually, the point cloud data for this frame will exhibit a certain degree of "deformation", failing to correspond to the environmental information accurately. This is known as LiDAR's motion distortion [8].

To correct the above motion distortion, we need to transform the coordinate systems of all points in the point cloud to a common one, such as the LiDAR coordinate system of the first point. Essentially, this process compensates for the motion of the LiDAR. Let p_i represent the coordinates of ith point in the LiDAR coordinate system at the current time, and the pose change from the coordinate system of p_i to that of the first point can be denoted as T_{1i}. Then, the coordinates of p_i in the LiDAR coordinate system of the first point can be expressed as

$$p_i^1 = T_{1i}p_i \quad (2.19)$$

The key is to determine T_{1i} for each point. In practice, the usual approach is to measure the motion information of the LiDAR, such as the pose change ΔT, which represents the transformation of the LiDAR's coordinate system between the start and end of a single point cloud frame captured over a time interval of Δt. Then, based on the time difference Δt_i between p_i and the starting, T_{1i} can be obtained through linear interpolation under the assumption of short-term constant velocity through the following formula:

$$T_{1i} = \frac{\Delta t_i}{\Delta t}\Delta T \quad (2.20)$$

The pose change ΔT can be obtained using pose information provided by an Inertial Navigation System (INS) or LiDAR odometry. If the pose change is calculated using an Inertial Measurement Unit (IMU), additional initial velocity information of the LiDAR or the vehicle is required. By transforming the coordinates of all points in a frame into the same coordinate system according to (2.19) and (2.20), the distortion correction process is completed.

2.4 MOTION ESTIMATION

Motion estimation [3] refers to the process of determining the relative pose change between two frames by analyzing observation data, which can include correspondences between 2D image features, 3D points, or a combination of both. According to the type of observation data used, motion estimation can be categorized into three classes [3,6]: 2D–2D correspondence-based epipolar geometry, 3D–2D correspondence-based Perspective-n-Point (PnP), and 3D–3D correspondence-based Iterative Closest Point (ICP).

2.4.1 Epipolar Geometry

Given two images and the matched 2D–2D feature points, the relative transformation R, t can be solved by utilizing epipolar constraint.

2.4.1.1 Epipolar Constraint

Figure 2.5 gives the illustration of the epipolar constraint. O_1 and O_2 are the optical centers of camera 1 and camera 2. Correspondingly, I_1 and I_2 are the images of cameras 1 and 2. The corresponding pixels of 3D point P in camera 1 and camera 2 are denoted as p_1 and p_2. The line O_1O_2 intersects plane I_1 and I_2 at points e_1 and e_2, where e_1, e_2 are called epipoles, and O_1O_2 is called baseline. Besides, e_2 is the corresponding pixel of O_1 in image I_2, and e_1 is the corresponding pixel of O_2 in image I_1. Assume that the coordinate of point P in the first frame is $P = \begin{bmatrix} X & Y & Z \end{bmatrix}^T$, and the rotation and translation transformations from the first frame to the second frame are R_{21} and t_{21}. According to Equation (2.15), we have

$$Z_1 p_1 = KP_1$$
$$Z_2 p_2 = K(R_{21}P_1 + t_{21}) \tag{2.21}$$

According to Equation (2.21), we get

$$Z_2 K^{-1} p_2 = R_{21} K^{-1} Z_1 p_1 + t_{21} \tag{2.22}$$

Let $x_2 = K^{-1}p_2$ and $x_1 = K^{-1}p_1$, and we get $x_1 = P_1 / Z_1$ and $x_2 = P_2 / Z_2$. Therefore, x_1 and x_2 can be regarded as the normalized coordinates of P_1 and P_2. Then, left multiply both sides by $\hat{t}_{21}$

$$Z_2 \hat{t}_{21} x_2 = Z_1 \hat{t}_{21} R_{21} x_1 \tag{2.23}$$

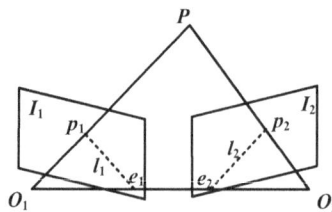

FIGURE 2.5 The epipolar constraint.

Left multiply both sides by x_2^T,

$$Z_2 x_2^T \hat{t}_{21} x_2 = Z_1 x_2^T \hat{t}_{21} R_{21} x_1 \tag{2.24}$$

Finally, we get

$$x_2^T \hat{t}_{21} R_{21} x_1 = 0 \tag{2.25}$$

$$p_2^T K^{-T} \hat{t}_{21} R_{21} K^{-1} p_1 = 0 \tag{2.26}$$

Both Equations (2.25) and (2.26) are termed as epipolar constraints. Define essential matrix $E_{21} = \hat{t}_{21} R_{21}$ and fundamental matrix $F_{21} = K^{-T} E_{21} K^{-1}$. The above epipolar constraints can be simplified as follows:

$$x_2^T E_{21} x_1 = p_2^T F_{21} p_1 = 0 \tag{2.27}$$

2.4.1.2 Pose Estimation

The epipolar constraint gives the spatial relationship between a pair of matched pixel points, where the essential matrix contain the relative pose R_{21} and t_{21}. Therefore, we can first estimate the essential matrix E_{21} by utilizing multiple pixel point correspondences. Then, the camera motion R_{21} and t_{21} can be figured out.

Define the normalized coordinate $x_1 = \begin{bmatrix} g_1, h_1, 1 \end{bmatrix}^T$ and $x_2 = \begin{bmatrix} g_2, h_2, 1 \end{bmatrix}^T$, according to Equation (2.27), we have

$$\begin{bmatrix} g_2, h_2, 1 \end{bmatrix} \begin{bmatrix} e_1 & e_2 & e_3 \\ e_4 & e_5 & e_6 \\ e_7 & e_8 & e_9 \end{bmatrix} \begin{bmatrix} g_1 \\ h_1 \\ 1 \end{bmatrix} = 0 \tag{2.28}$$

Let $e = \begin{bmatrix} e_1, e_2, e_3, e_4, e_5, e_6, e_7, e_8, e_9 \end{bmatrix}^T$, Equation (2.28) can be rewritten as follows:

$$\begin{bmatrix} g_2 g_1, g_2 h_1, g_2, h_2 g_1, h_2 h_1, h_2, g_1, h_1, 1 \end{bmatrix} e = 0 \tag{2.29}$$

Considering n pairs of feature points between two images, we get the following linear equation:

$$\begin{bmatrix} g_2^1 g_1^1 & g_2^1 h_1^1 & g_2^1 & h_2^1 g_1^1 & h_2^1 h_1^1 & h_2^1 & g_1^1 & h_1^1 & 1 \\ g_2^2 g_1^2 & g_2^2 h_1^2 & g_2^2 & h_2^2 g_1^2 & h_2^2 h_1^2 & h_2^2 & g_1^2 & h_1^2 & 1 \\ \vdots & \vdots & \vdots & \vdots & \vdots & \vdots & \vdots & \vdots & \vdots \\ g_2^n g_1^n & g_2^n h_1^n & g_2^n & h_2^n g_1^n & h_2^n h_1^n & h_2^n & g_1^n & h_1^n & 1 \end{bmatrix} e = 0 \tag{2.30}$$

Essential matrix E_{21} contains rotation and translation components, they corresponds to 6 degrees of freedom. Considering the equivalence of scales, E_{21} has 5 degrees of freedom. Therefore, we can use at least five pairs of points to solve it [9]. Readers can refer to the five-point algorithm [10] and eight-point algorithm [11] to get the essential matrix.

Next, we restore the relative pose R_{21} and t_{21} from the estimated essential matrix E_{21}. Executing singular value decomposition (SVD) [12] on E_{21}, and then $E_{21} = U\Sigma V^T$. U, V are orthogonal matrixes, and Σ is the singular value matrix. Since $E_{21} = t_{21}^{\wedge} R_{21}$, two groups of R_{21}, t_{21} are obtained:

$$t_{21}^1{}^{\wedge} = UR_z\left(\frac{\pi}{2}\right)\Sigma U^T \qquad R_{21}^1 = UR_z^T\left(\frac{\pi}{2}\right)V^T$$

$$t_{21}^2{}^{\wedge} = UR_z\left(-\frac{\pi}{2}\right)\Sigma U^T \qquad R_{21}^2 = UR_z^T\left(-\frac{\pi}{2}\right)V^T \tag{2.31}$$

where $R_z\left(\frac{\pi}{2}\right)$ represents the rotation matrix obtained by rotating 90 degrees along the z axis. Due to the equivalence of E_{21} and $-E_{21}$, imposing minus sign on the above $t_{21}^1{}^{\wedge}$, $t_{21}^2{}^{\wedge}$ is also feasible. Therefore, there exist four group of solutions of R_{21}, t_{21}. In practice, for each group of solution, we can compute the corresponding 3D coordinate based on the matched feature points on both images and then two depths in cameras 1 and 2 are further figured out. The only R_{21}, t_{21} with two positive depths is regarded as the final result.

2.4.2 PnP

The PnP problem involves estimating the camera pose $\{R,t\}$ from a set of 3D points in space and their corresponding 2D projections in an image. Over the years, various methods have been developed to solve the PnP problem, including direct linear transformation (DLT) [13], efficient PnP (EPnP) [14], and nonlinear optimization techniques. Among these, DLT stands out as one of the simplest and most straightforward approaches, making it a common starting point for understanding PnP solutions.

For a 3D point $P_i = (X_i, Y_i, Z_i)$, and its corresponding 2D projection points with homogeneous coordinates $p_i = \begin{bmatrix} u_i & v_i & 1 \end{bmatrix}^T$, the following equation can be obtained according to (2.16):

$$Z_i\begin{bmatrix} u_i \\ v_i \\ 1 \end{bmatrix} = K[R|t]\begin{bmatrix} X_i \\ Y_i \\ Z_i \\ 1 \end{bmatrix} \tag{2.32}$$

where $[R|t]$ is a 3×4 matrix composed of the elements of unknown rotation matrix R and translation vector t. We can define an unknown 3×4 matrix A $= [R|t]$ = $\begin{bmatrix} a_{11} & a_{12} & a_{13} & a_{14} \\ a_{21} & a_{22} & a_{23} & a_{24} \\ a_{31} & a_{32} & a_{33} & a_{34} \end{bmatrix}$. Then, (2.32) can be written as follows:

$$
Z_i x_i = \begin{bmatrix} a_{11} & a_{12} & a_{13} & a_{14} \\ a_{21} & a_{22} & a_{23} & a_{24} \\ a_{31} & a_{32} & a_{33} & a_{34} \end{bmatrix} \overline{P_i} \tag{2.33}
$$

where $\overline{P_i} = [X_i, Y_i, Z_i, 1]^T$ and $x_i = K^{-1} p_i$ are the normalized coordinates of P_i. Based on the derivation in Section 2.4.1, $x_i = \dfrac{P_i}{Z_i} = [g_i, h_i, 1]^T$. Denote $a_1 = [a_{11}, a_{12}, a_{13}, a_{14}]^T$, $a_2 = [a_{21}, a_{22}, a_{23}, a_{24}]^T$, $a_3 = (a_{31}, a_{32}, a_{33}, a_{34})^T$, we can obtain that

$$
g_i = \frac{a_1^T \overline{P_i}}{a_3^T \overline{P_i}}, \quad h_i = \frac{a_2^T \overline{P_i}}{a_3^T \overline{P_i}} \tag{2.34}
$$

The goal is to determine the camera pose, encapsulated within matrix A. As indicated by Equation (2.34), each pair of points imposes two constraints. Given that matrix A comprises 12 elements, it can be linearly resolved by employing at least six pairs of points. This is achieved by accumulating the respective constraints for each pair:

$$
\begin{bmatrix} \overline{P_1}^T & 0 & -g_1 \overline{P_1}^T \\ 0 & \overline{P_1}^T & -h_1 \overline{P_1}^T \\ & \vdots & \\ \overline{P_6}^T & 0 & -g_6 \overline{P_6}^T \\ 0 & \overline{P_6}^T & -h_6 \overline{P_6}^T \end{bmatrix} \begin{bmatrix} a_1 \\ a_2 \\ a_3 \end{bmatrix} = 0 \tag{2.35}
$$

It is important to recognize that the first three columns of matrix A constitute the rotation matrix R, which is inherently an orthonormal matrix. Nonetheless, the solution we derive by DLT may not precisely adhere to this attribute. Fortunately, reference [3] indicates that we can project solution A onto the SE(3) manifold by QR factorization to obtain the final rotation and translation components.

2.4.3 ICP

ICP is a solution to estimate the pose based on two sets of matched 3D points, $P = \{p_1, p_2, ..., p_n\}$, $P' = \{p_1', p_2', ..., p_n'\}$. The objective is to find a transformation $\{R^*, t^*\}$ such that

$$
R^*, t^* = \arg\min_{R,t} \frac{1}{2} \sum_{i=1}^{n} \left\| p_i - (R p_i' + t) \right\|_2^2 \tag{2.36}
$$

where R is the rotation matrix and t is the translation vector. Many methods have been proposed by researchers to solve the ICP problem, including techniques such as SVD, non-linear optimization, and others. Notably, it has been proven that the ICP problem for two sets of matched 3D points has a closed-form solution using SVD, making it both efficient and mathematically robust for estimating the optimal transformation.

Define the mass-center of P and P' as p and p', such that

$$p = \frac{1}{n}\sum_{i=1}^{n}p_i \qquad p' = \frac{1}{n}\sum_{i=1}^{n}p'_i \tag{2.37}$$

The error term in (2.36) can be transformed as follows:

$$\frac{1}{2}\sum_{i=1}^{n}\left\|p_i - \left(Rp'_i + t\right)\right\|_2^2 = \frac{1}{2}\sum_{i=1}^{n}\left\|p_i - Rp'_i - t - p + Rp' + p - Rp'\right\|_2^2$$

$$= \frac{1}{2}\sum_{i=1}^{n}\left(\begin{array}{c}\left\|p_i - p - R\left(p'_i - p'\right)\right\|_2^2 + \left\|p - Rp' - t\right\|_2^2 \\ +2\left(p_i - p - R\left(p'_i - p'\right)\right)^T\left(p - Rp' - t\right)\end{array}\right) \tag{2.38}$$

$$= \frac{1}{2}\sum_{i=1}^{n}\left(\left\|p_i - p - R\left(p'_i - p'\right)\right\|_2^2 + \left\|p - Rp' - t\right\|_2^2\right)$$

The last step of the above equation is because $p - Rp' - t$ does not vary with i and $\sum_{i=1}^{n}\left(p_i - p - R\left(p'_i - p'\right)\right) = 0$. Thus, Equation (2.36) can be transformed to

$$R^*, t^* = \arg\min_{R,t}\frac{1}{2}\sum_{i=1}^{n}\left(\left\|p_i - p - R\left(p'_i - p'\right)\right\|_2^2 + \left\|p - Rp' - t\right\|_2^2\right) \tag{2.39}$$

The first term of Equation (2.39) is only dependent on R, so we can first minimize this term to solve for the optimal rotation matrix R^*. Then, by setting the second term to zero, we substitute R with R^* to solve for the optimal translation vector t^*. For the first term, we define $q_i = p_i - p$, $q'_i = p'_i - p'$. Then, we have

$$\frac{1}{2}\sum_{i=1}^{n}\left\|q_i - Rq'_i\right\|_2^2 = \frac{1}{2}\sum_{i=1}^{n}-2q_i^T Rq'_i + q_i^T q_i + q'^T_i R^T Rq'_i$$

$$= \frac{1}{2}\sum_{i=1}^{n}-2q_i^T Rq'_i + \text{const} = -\text{tr}\left(R\sum_{i=1}^{n}q'_i q_i^T\right) + \text{const} \tag{2.40}$$

Following [15], we define a 3×3 matrix $W = \sum_{i=1}^{n}q_i q'^T_i$. Through SVD decomposition, we have $W = U\Sigma V^T$, where U and V are orthonormal matrices, and Σ is a diagonal matrix of singular values, with diagonal elements arranged from the largest to smallest. If W is full-rank, R can be solved:

$$R^* = UV^T \tag{2.41}$$

From the last term of Equation (2.39), we can obtain t^* as follows:

$$t^* = p - R^* p' \tag{2.42}$$

2.5 CONCLUSION

In this chapter, we have established the essential mathematical foundation and theoretical framework for localization and mapping, which are critical for the autonomous movement and navigation of mobile robots. Starting with the representation of 3D rigid body motion, we explored various representation methods and their conversions, providing the basis for understanding robot motion. We then introduced sensor models, including the pinhole camera model and LiDAR model, which form the core of associating 3D objects with their corresponding observations. Building on this, we examined key motion estimation approaches, highlighting how sensor observations and their correspondences can be utilized to compute the robot's pose transformation.

REFERENCES

[1] Wikipedia. https://en.wikipedia.org/wiki/Rotation_matrix.
[2] Wikipedia. https://en.wikipedia.org/wiki/Axis%E2%80%93angle_representation.
[3] Gao, X., Zhang, T., Yan, Q., & Liu, Y. (2017). *14 Lectures on Visual SLAM: From Theory to Practice*. Publishing House of Electronics Industry (In Chinese).
[4] Wikipedia. https://en.wikipedia.org/wiki/Rodrigues%27_rotation_formula.
[5] Wikipedia. https://en.wikipedia.org/wiki/Pinhole_camera_model.
[6] Hartley, R., & Zisserman, A. (2003). *Multiple View Geometry in Computer Vision*. Cambridge University Press.
[7] Barfoot, T. D. (2024). *State Estimation for Robotics*. Cambridge University Press.
[8] Liang, S., Cao, Z., Yu, J., Tan, M., & Wang, S. (2022). A tight filtering and smoothing fusion method with feature tracking for LiDAR odometry. *IEEE Sensors Journal*, 22(13), 13622–13631.
[9] Li, H., & Hartley, R. (2006). Five-point motion estimation made easy. In *Proceedings of International Conference on Pattern Recognition*, Hong Kong, China (pp. 630–633).
[10] Nistér, D. (2004). An efficient solution to the five-point relative pose problem. *IEEE Transactions on Pattern Analysis and Machine Intelligence*, 26(6), 756–770.
[11] Hartley, R. I. (1997). In defense of the eight-point algorithm. *IEEE Transactions on Pattern Analysis and Machine Intelligence*, 19(6), 580–593.
[12] Wikipedia. https://en.wikipedia.org/wiki/Singular_value_decomposition.
[13] Chen, L., Armstrong, C. W., & Raftopoulos, D. D. (1994). An investigation on the accuracy of three-dimensional space reconstruction using the direct linear transformation technique. *Journal of Biomechanics*, 27(4), 493–500.
[14] Lepetit, V., Moreno-Noguer, F., & Fua, P. (2009). EPnP: An accurate $O(n)$ solution to the PnP problem. *International Journal of Computer Vision*, 81, 155–166.
[15] Arun, K. S., Huang, T. S., & Blostein, S. D. (1987). Least-squares fitting of two 3-D point sets. *IEEE Transactions on Pattern Analysis and Machine Intelligence*, 9(5), 698–700.

CHAPTER 3

Real-Time Semantic Visual SLAM with Points and Objects

3.1 INTRODUCTION

Compared to the direct visual SLAM methods [1,2], feature-based approaches [3–5] rely on a sparse set of distinctive features, reducing computational complexity. However, different viewpoints and illuminations can lead to variations in local appearance and brightness of the same point, which will cause the tracking failure of points with incorrect data association. Thus, the localization accuracy of the visual SLAM is decreased. Besides, these methods mainly focus on low-level geometric information, which possibly results in a weak interaction with complex surrounding environments [6]. With the development of deep learning, great progresses have been made in object detection and object segmentation, whose high-level semantic information can better adapt to viewpoint and illumination changes. The purpose of object detection is to infer the locations and class labels of objects, where the location of the object is represented in the form of a bounding box. For object detection based on deep learning, it can be classified into approaches based on regional proposal and without regional proposal. The former is a two-stage process: firstly, generate a series of candidate regions and then extract the features of the candidate regions for classification and boundary regression. Its popular methods include regions with convolutional neural network features (R-CNN) [7], Fast R-CNN [8], Faster R-CNN [9], and so on. For the approaches without regional proposal, the global information of the image is directly used, and You Only Look Once (YOLO) [10], YOLO9000 [11], YOLOv3 [12], and single-shot multibox detector [13] are the representative methods. Different from the bounding box of object detection, object segmentation predicts the class labels pixel by pixel, and it is related to semantic segmentation [14,15] and instance segmentation [16]. A possible problem of object segmentation is its computational cost, which makes it hard to integrate into a real-time SLAM.

Driven by object detection and segmentation based on deep learning, the researchers are concerned with semantic visual SLAM with the combination of object detection

DOI: 10.1201/9781003643630-3

and segmentation. Semantics can not only help SLAM achieve better localization [17–20] but also establish more abundant maps. To improve the localization accuracy, semantic constraints are added. Lianos et al. constructed a semantic error function by utilizing semantic segmentation to promote point–point association [17]. An et al. evaluated the importance of each semantic category based on semantic segmentation for better visual features and the removal of outliers in the matching process [18]. On this basis, the accuracy and robustness of localization are improved. Besides semantic constraint, pose optimization of objects is also considered. A 3D cuboid object detection approach is proposed [19], and it is combined with the Oriented FAST and Rotated BRIEF (ORB) feature points to, respectively, build semantic error functions for static and dynamic environments. On this basis, poses of points, 3D cuboids, and cameras are jointly optimized. Similarly, Li et al. utilized 3D object detection with viewpoint classification as well as feature points for constructing semantic constraints [20], which is suitable for both static and dynamic conditions.

It should be noted that existing semantic SLAM approaches mainly concern the constraints of camera-landmark, camera-camera, as well as different types of landmarks, where a landmark can be a point-type, and it can also be an object type. The constraint of landmarks with the same kind is seldom considered. In fact, there exists invariance in terms of relative distance and orientation between two static object landmarks, and it may be changed if only the aforementioned constraints are employed. It is beneficial for the localization by introducing the relative constraints among objects into the SLAM optimization process. In this chapter, we propose a real-time visual Point–Object SLAM (PO-SLAM) approach on the basis of RGBD ORB-SLAM2, which incorporates object–object constraint in the bundle adjustment (BA) optimization process. To ensure the real-time performance of the system while considering the instantiation of the objects, YOLOv3 [12] is adopted, and it is combined with a rough geometric segmentation based on depth histogram to obtain the contours of objects, which can improve the association quality. Moreover, the object–object constraint is reflected by the relative position invariance of objects, which is converted to the length and orientation invariances of the line segment connecting every two objects in each frame. This provides additional information for pose optimization.

In the following, we will describe the proposed PO-SLAM approach combining points and objects in detail. Then, the experiments are presented, and finally, we conclude the chapter.

3.2 THE PROPOSED SEMANTIC PO-SLAM WITH POINTS AND OBJECTS

The framework of the proposed semantic PO-SLAM is shown in Figure 3.1, where point features, point–point association, and point–point constraint are directly used according to ORB-SLAM2 [5]. In the feature extraction module, object features are extracted from the color image provided by RGBD camera using YOLOv3 [12]. Considering that object detection cannot accurately express the contours of objects, we utilize the depth image to geometrically segment the detected objects based on depth histograms. Then, combined with point features, point–object association is executed to obtain the feature points on each detected object. After extracting the features of every frame, we track the features

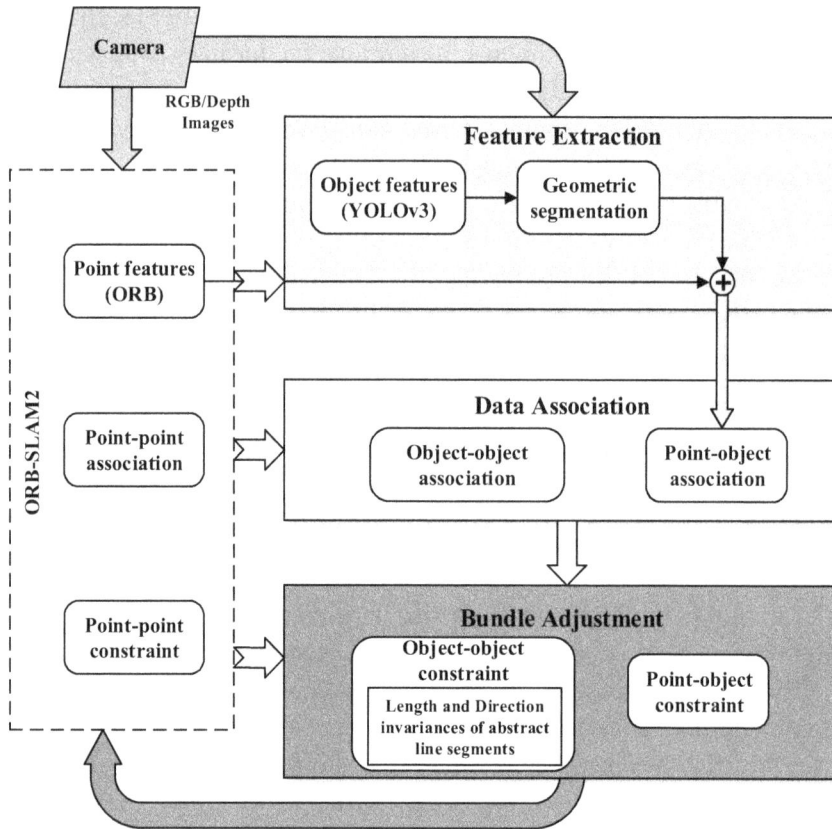

FIGURE 3.1 Overall framework of the semantic PO-SLAM approach.

between the current frame and the previous frame. Besides the point–point association, object–object inter-frame association is also executed. On this basis, the extracted point and object features as well as the association results, are involved in the BA optimization process. With the help of loop closing of ORB-SLAM2, SLAM is finally implemented. In the following, we will address the PO-SLAM in detail.

3.2.1 Feature Extraction

Low-level point features are combined with high-level semantic object features in our SLAM. The reader may refer to the study by Mur-Artal and Tardós [5] for point feature extraction, and in this section, we focus on the extraction of object features.

3.2.1.1 Object Features Extraction

Object features including the objects number, categories, and the positions are favorable for data association of SLAM due to the reliability of high-level feature. In this chapter, YOLOv3 [12] is utilized to detect the objects at each frame, where the deep network is trained on the MS COCO data set including 80 categories of common objects. By object detection, the bounding boxes, labels, and label confidences of objects are obtained. Note that we only reserve the results with confidence of more than 70%.

3.2.1.2 Geometric Segmentation

For object detection, the resulting bounding box surrounding an object cannot fit the actual boundary of object completely, and some background information is inevitably contained. In this case, it is not easy to judge whether a feature point is on an object, which will affect the determination of the object's position. Also, in spite of good performance in segmentation effect, instance segmentation based on deep learning needs to take more time. A fast segmentation solution to extract the foreground in the bounding box of an object is required. Herein, a geometric segmentation based on depth histogram is presented.

In a detection bounding box, there are only two types of pixels: background and foreground. Their differentiation may be solved using depth information that reflects the distance between an object and the camera, and a depth threshold to separate the foreground from the background needs to be determined. With the depth values of foreground and background, we utilize the Otsu threshold segmentation method [21] to segment the depth values by maximizing the interclass variance of these two parts. Otsu is a method to automatically determine the threshold; however, it is sensitive to noise. For the depth map provided by the RGBD camera, there exists the case where the depth value of a pixel is 0, which may be caused by the pixels outside the depth range or miss detection. Those pixels with a depth value of 0 in the depth map should be first filtered out before calculating the depth threshold.

To obtain the geometrical segmentation for an object, the depth image of current frame is cropped according to the predicted bounding box, and one can obtain the depth submap $d_{h \times w}$. After $d_{h \times w}$ is filtered, its values are scaled to [0, 255], which is used to acquire the threshold using the Otsu method. On this basis, the foreground and background corresponding to the object are separated. The detailed process is given in Algorithm 3.1. $\left(b^l, b^b\right)$ and $\left(b^r, b^t\right)$ are the left bottom coordinate and the upper right coordinate of the bounding box, respectively, and $h = b^b - b^t$, $w = b^r - b^l$. D_{th} refers to the depth threshold, and the segmentation mask is labeled as M.

Algorithm 3.1 Geometrical Segmentation Process

Input: bounding box $o = \left\{\left(b^l, b^b\right), \left(b^r, b^t\right)\right\}$ of detected object, depth image $D_{m \times n}$ of current frame.

Output: segmentation mask $M_{h \times w}$ for the detected object:

1 $d_{h \times w} = D(o)$, vector_$d = []$, $D_{th} = 0$;
2 vector_$d_{max} = 0$, vector_$d_{min} = 0$;
3 **for** $i = 0, 1, \ldots, h - 1$
4   **for** $j = 0, 1, \ldots, w - 1$
5   **if** $d[i][j] > 0$ **then**
6   vector_d.push_back$\left(d[i][j]\right)$;
7   **end if**

8   **end for**

9 **end for**

10 $\text{vector\_}d_{\max} = \max(\text{vector\_}d);$

11 $\text{vector\_}d_{\min} = \min(\text{vector\_}d);$

12 **for** $k = 0, 1, \ldots, \text{Size}(\text{vector\_}d) - 1$ **do**

13    $\text{vector\_}d[k] = \dfrac{\text{vector\_}d[k] - \text{vector\_}d_{\min}}{\text{vector\_}d_{\max} - \text{vector\_}d_{\min}} \times 255;$

14 **end for**

15 $D_{\text{th}} = \text{Otsu\_threshold}(\text{vector\_}d);$

16 $D_{\text{th}} = D_{\text{th}} \times (\text{vector\_}d_{\max} - \text{vector\_}d_{\min})/255 + \text{vector\_}d_{\min};$

17 **for** $i = 0, 1, \ldots, h - 1$

18   **for** $j = 0, 1, \ldots, w - 1$

19   **if** $d[i][j] > 0$ & $d[i][j] < D_{\text{th}}$ **then**

20        $M[i][j] = 1;$

21   **else**

22        $M[i][j] = 0;$

23   **end if**

24   **end for**

25 **end for**

26 return M

Figure 3.2 illustrates the segmentation result. Take the teddy bear in the original image from the TUM data set [22] (see Figure 3.2a) as an example. Figure 3.2b provides the detection result, and the depth histogram of the pixels in the bounding box is presented in Figure 3.2c. One can see that the depth values are divided into two parts by the dashed line corresponding to the depth threshold D_{th}. The left and right parts of the dashed line correspond to the depth values of foreground and background. With the segmentation, the extracted foreground is given in Figure 3.2d for the object in Figure 3.2b.

3.2.2 Data Association

As a reflection of the common view between frames, data association is important in solving camera poses and landmark positions of SLAM. In addition to the association of inter-frame point features used in ORB-SLAM2 [5], we also take the correlation of point features and object features in each frame as well as the association of inter-frame object features into account.

3.2.2.1 Point–Object Association

As mentioned above, for each detected bounding box in each frame, the foreground image is separated by the depth image information, and the feature points located in the foreground area are used as the feature points corresponding to the object. The association of points and objects is used to calculate the point–object error in the subsequent BA optimization. Figure 3.3 gives an illustration of feature points and detected objects for a selected

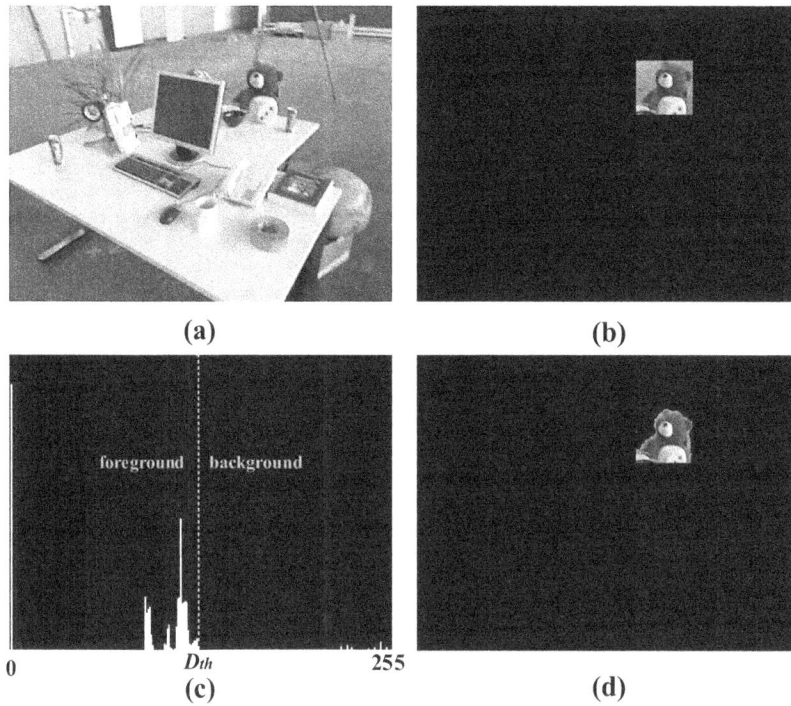

FIGURE 3.2 The geometric segmentation. (a) Original image. (b) One detected object. (c) The depth histogram of the bounding box in (b). (d) The extracted foreground after the segmentation.

FIGURE 3.3 Illustration of feature points and detected objects for an image in fr2/desk of TUM RGBD dataset.

image in fr2/desk of TUM RGBD dataset [22]. Note that multiple object instances of the same class can be distinguished by the positions of their bounding boxes, and the points which do not belong to any detected object are considered as the background. When the points fall within the bounding box of an object and their classes match the class of the bounding box, they are regarded as the feature points associated with the object.

3.2.2.2 Object–Object Association

Object–object association between two frames is similar to standard object tracking. Since we have known the categories of the objects in each frame, we can only be concerned with the object categories that simultaneously appear in two frames. At first, the center (u_c, v_c) of an object in the previous color image is un-projected to the world coordinate system by its depth d_c and the camera pose $T_{cw,\text{pre}}$ of previous frame. Then 3D position P_c of the object center is projected to the current image using the camera pose $T_{cw,\text{cur}}$ of the current frame:

$$P_c = \pi^{-1}\left(T_{cw,\text{pre}}, d_c, u_c, v_c\right) \tag{3.1}$$

$$(u_c, v_c)_{\text{proj}} = \pi\left(T_{cw,\text{cur}}, P_c\right) \tag{3.2}$$

where π and π^{-1} represent the projection from 3D space to 2D image and un-projection from 2D image to 3D space, respectively. $(u_c, v_c)_{\text{proj}}$ refers to the projection of P_c on the current frame. After the projection of the object center on the current frame is acquired, we check the relationship of the projection and the bounding boxes in the current frame with the constraint of the same object label. If the distance between the projection and the center of a bounding box in the current frame is less than a given threshold, the corresponding two objects are considered as a successful match.

3.2.3 Bundle Adjustment

Combining point and object features, constraints with geometric and semantic relationships are constructed to optimize camera poses and 3D point positions. The sets of image sequence, positions of 3D points, and objects in the world coordinate system are denoted as $I = \{I_k\}$, $P = \{P_i\}$, $O = \{O_j\}$, respectively, where k, i, and j are their corresponding indexes. For a 3D point, it is either on the object or belongs to the background. We label the position of the ith point on the jth object as ^jP_i. Also, the object is represented by the points inside the object and its position, and $\{O_j\} = \left\{\left\{{}^jP_i, C_j\right\}\right\}$, where C_j is the 3D position of the jth object.

   We can observe the measurements corresponding to 3D points and objects from each frame. $o = \{o_{kj}\}$ and $z = \{z_{ki}\}$ are used to stand for the observations of jth object O_j and the ith point in the kth frame. $o_{kj} = \left\{b_{kj}^l, b_{kj}^b, b_{kj}^r, b_{kj}^t, c_{kj}, l_{kj}\right\}$, where c_{kj}, l_{kj} are the observations of object position and class label. We denote $^jz_{ki}$ with the observation on the jth object for the ith point in the kth frame.

3.2.3.1 BA Formulation

Our semantic optimization process can be described as the following problem: given the observations $\{z_{ki}\}$ of points in the kth frame and observations $\{o_{kj}\}$ of objects $\{O_j\}$ in the kth frame, find the optimized camera pose $T^*_{cw,k}$ and the positions $\{P^*_i\}$ of the points, where $T_{cw,k} \in \text{SE}(3)$ is used to convert 3D points from the world coordinate system to the camera coordinate system, and $P_i \in R^3$. In the BA process, the optimization process is executed by minimizing the errors between the predicted values and the measured values, which is a nonlinear least squares problem. Our measurement errors consist of point–point error, point–object error, and object–object error, and the optimization function can be formulated as follows:

$$T^*_{cw,k}, P^*_i = \arg \min_{T_{cw,k}, P_i} \sum_{k, j, i} \left\{ \left\| e_{pp}\left(z_{ki}, T_{cw,k}, P_i\right) \right\|^2_{\Sigma_{k,i}} \right.$$
$$\left. + \left\| e_{oo}\left(o_{kj_1}, o_{kj_2}, T_{cw,k}, O_{j_1}, O_{j_2}\right) \right\|^2_{\Sigma_{k,j1,j2}} + \left\| e_{po}\left(o_{kj}, T_{cw,k}, {}^j P_i\right) \right\|^2_{\Sigma_{k,i,j}} \right\} \tag{3.3}$$

where $e_{pp}(\cdot)$, $e_{po}(\cdot)$ and $e_{oo}(\cdot)$ represent the errors between projected point on the image by the camera pose and observation point for P_i, between projected point and 2D bounding box, and between two objects, respectively. In this chapter, the Levenberg–Marquardt method is adopted to solve this problem.

3.2.3.2 Error Functions

Point–point error: With the ORB features, point–point error (i.e., re-projection error) is given by [5]

$$e_{pp}\left(z_{ki}, T_{cw,k}, P_i\right) = \pi\left(T_{cw,k}, P_i\right) - z_{ki} \tag{3.4}$$

Point–object error: Based on the point and object data association, we get $\{{}^j P_i\}$ that belong to an object O_j. Theoretically, after these points are projected into the current frame, they should fall into the corresponding 2D bounding box of the object O_j, but that is not always the case. Our point–object error for the point ${}^j P_i$ is as follows:

$$e_{po}\left(o_{kj}, T_{cw,k}, {}^j P_i\right) = (\text{err}_u, \text{err}_v) \tag{3.5}$$

where

$$\text{err}_u = \begin{cases} 0 & b^l_{kj} \le u_{\text{proj}} \le b^r_{kj} \\ u_{\text{proj}} - b^r_{kj} & u_{\text{proj}} > b^r_{kj} \\ b^l_{kj} - u_{\text{proj}} & u_{\text{proj}} < b^l_{kj} \end{cases} \tag{3.6}$$

$$\text{err}_v = \begin{cases} 0 & b_{kj}^t \le v_{\text{proj}} \le b_{kj}^b \\ v_{\text{proj}} - b_{kj}^b & v_{\text{proj}} > b_{kj}^b \\ b_{kj}^t - v_{\text{proj}} & v_{\text{proj}} < b_{kj}^t \end{cases} \tag{3.7}$$

$$\left(u_{\text{proj}}, v_{\text{proj}} \right) = \pi \left(T_{cw,k}, {}^j P_i \right) \tag{3.8}$$

$\left(u_{\text{proj}}, v_{\text{proj}} \right)$ is the projected pixel coordinate of ${}^j P_i$, and err_u and err_v are the u-axis error and v-axis error between projected point and 2D bounding box. It should be noted that when the projection point is inside the detected bounding box, the cost function is always zero, and thus this constraint is relatively coarse. Only when the projection point falls outside the detection box does the penalty take effect.

Object–object error: we acquire the feature points belonging to the objects as well as corresponding 3D points through the point–object data association. We then use the coordinate centroid of these 3D points as the 3D position of the object, with the coordinate centroid of corresponding ORB feature points in the image as the observation of the object position, which is described as

$$C_j = \frac{1}{N} \sum_i {}^j P_i \tag{3.9}$$

$$c_{kj} = \frac{1}{N} \sum_i {}^j z_{ki} \tag{3.10}$$

where N is the number of points anchored to the object O_j.

The relative position between two objects is constrained by distance and orientation. In order to solve the problem, we connect the positions of two objects into an abstract line segment, and thus, the distance and direction constraints can be converted to the invariance of length and direction of the line segment. We define c_{kj_1} and c_{kj_2} as the observations of positions for objects O_{j_1} and O_{j_2} in the kth frame, respectively. Correspondingly, C_{j_1} and C_{j_2} represent the 3D positions of objects O_{j_1} and O_{j_2}. According to [23], we define $c_{kj_1}^h$ and $c_{kj_2}^h$ as the homogeneous coordinates of c_{kj_1} and c_{kj_2} for the parameterized representation of the line segment. Thus, the line through c_{kj_1} and c_{kj_2} can be expressed as

$$l = \frac{c_{kj_1}^h \times c_{kj_2}^h}{\left| c_{kj_1}^h \times c_{kj_2}^h \right|} \tag{3.11}$$

According to the direction invariance constraint, we can infer that the projection points of object O_{j_1} and object O_{j_2} should be located on the line l. The direction error can be denoted as

$$e_{oo\_\text{dir}} = \left(l^T \pi \left(T_{cw,k}, C_{j_1} \right), l^T \pi \left(T_{cw,k}, C_{j_2} \right) \right) \tag{3.12}$$

The length invariance of the line segment indicates that the distance between the projected points is the same as that of c_{kj_1} and c_{kj_2}. Then the distance error is given by

$$e_{oo\_dis} = D\left(p_1, p_2\right) - D(c_{kj_1}, c_{kj_2}) = \sqrt{\left(p_1^u - p_2^u\right)^2 + \left(p_1^v - p_2^v\right)^2}$$

$$-\sqrt{\left(c_{kj_1}^u - c_{kj_2}^u\right)^2 + \left(c_{kj_1}^v - c_{kj_2}^v\right)^2} \tag{3.13}$$

where

$$p_1 = \pi\left(T_{cw,k}, C_{j_1}\right) \tag{3.14}$$

$$p_2 = \pi\left(T_{cw,k}, C_{j_2}\right) \tag{3.15}$$

where p_1 and p_2 respectively represent the 2D pixel coordinates of the projections of C_{j_1} and C_{j_2} in the image and $D(\cdot)$ refers to the Euclidean distance of two pixels. e_{oo_dir} and e_{oo_dis} constitute the object–object error function.

3.3 EXPERIMENTS

In this section, we will evaluate the localization performance of our approach, and conduct the comparison with ORB-SLAM2.

3.3.1 Experimental Setup

We adopt the TUM RGBD SLAM data set and benchmark [22,24] to test and validate the approach. TUM data set consists of different types of sequences, which provide color and depth images with a resolution of 640×480 using a Microsoft Kinect sensor. YOLOv3 scales the original images to 416×416. Combining objects we concerned, such as book, keyboard, mouse, TV-monitor, cup, cell phone, remote, bottle, teddy bear, and potted plants, 10 sequences related to office environments are selected.

We adopt the following evaluation metrics [24]: absolute trajectory error with root mean square (ATE) and mean relative pose error (RPE), where ATE quantifies the difference between points of the estimated trajectory and their ground truths, whereas RPE assesses the local accuracy of the estimated poses in a fixed interval. All of the experiments are repeated five times and the median of these five results is considered as the final result. To clearly demonstrate the improvement of our method, ATE_{rel} [17] and RPE_{rel} are considered, where the former refers to the relative ATE and the latter reflects the relative RPE. $ATE_{rel} = \left(ATE_{ORB\_SLAM2} - ATE_{+semantics}\right) \times 100 / ATE_{ORB\_SLAM2}$ and $RPE_{rel} = \left(RPE_{ORB\_SLAM2} - RPE_{+semantics}\right) \times 100 / RPE_{ORB\_SLAM2}$, where $ATE_{+semantics}$ and $RPE_{+semantics}$ represent the ATE and RPE of our semantic SLAM. All of our experiments are run on a desktop with an Intel Core i7-7700HQ CPU and Nvidia GTX 1080 GPU. Only the object detection is executed on the GPU.

3.3.2 Experiment on the TUM RGBD Dataset

Tables 3.1 and 3.2 give the comparison of our PO-SLAM and ORB-SLAM2 over 10 sequences. To better evaluate our approach, we also consider two other methods: PO-SLAM1 and PO-SLAM2. These two methods correspond to the cases of PO-SLAM without point–object error in (3) and PO-SLAM without object–object error in (3), respectively. Note that the first seven sequences describe static scenes, whereas the last three sequences are related to dynamic scenes.

As can be seen in Tables 3.1 and 3.2, our PO-SLAM has an improvement of up to 10.46% in ATE and up to 10.95% in RPE compared with ORB-SLAM2. Overall, our three methods perform better than ORB-SLAM2 in both ATE and RPE for most of the sequences, and PO-SLAM performs best.

Figure 3.4 depicts the comparison of the trajectories obtained by PO-SLAM and ORB-SLAM2 on four sequences with the ground truth. It is seen that our trajectories are closer to the ground truth than ORB-SLAM2. Note that all ORB features extracted by ORB-SLAM2 are used in our point–point error. From Tables 3.1 and 3.2, our method has

TABLE 3.1 Comparison of Our Methods with ORB-SLAM2 According to Absolute Trajectory Errors

	ORB-SLAM2	PO-SLAM1		PO-SLAM2		PO-SLAM	
Sequence	ATM (m)	ATM (m)	ATE_{rel} (%)	ATM (m)	ATE_{rel} (%)	ATM (m)	ATE_{rel} (%)
fr1/desk	0.0153	0.0158	−3.27	0.0152	0.65	0.0153	0.00
fr1/desk2	0.0239	0.0220	7.95	0.0234	2.09	0.0214	10.46
Fr1/room	0.0537	0.0516	3.91	0.0484	9.87	0.0481	10.43
fr1/xyz	0.0097	0.0097	0.00	0.0096	1.03	0.0096	1.03
fr2/desk	0.0094	0.0092	2.13	0.0091	3.19	0.0090	4.26
fr2/xyz	0.0036	0.0037	−2.78	0.0036	0.00	0.0035	2.78
fr3/long_office	0.0100	0.0097	3.00	0.0102	−2.00	0.0096	4.00
fr3/sitting_xyz	0.0093	0.0090	3.23	0.0091	2.15	0.0091	2.15
fr3/sitting_static	0.0087	0.0080	8.05	0.0084	3.45	0.0079	9.20
fr3/walking_xyz	0.7127	0.7012	1.61	0.7128	−0.01	0.7025	1.43

TABLE 3.2 Comparison of Our Methods with ORB-SLAM2 According to Relative Pose Errors

	ORB-SLAM2	PO-SLAM1		PO-SLAM2		PO-SLAM	
Sequence	RPE (m)	RPE (m)	RPE_{rel} (%)	RPE (m)	RPE_{rel} (%)	RPE (m)	RPE_{rel} (%)
fr1/desk	0.0279	0.0283	−1.43	0.0271	2.87	0.0277	0.72
fr1/desk2	0.0409	0.0408	0.24	0.0399	2.44	0.0388	5.13
fr1/room	0.0840	0.0763	9.17	0.0762	9.29	0.0748	10.95
fr1/xyz	0.0129	0.0128	0.78	0.0127	1.55	0.0129	0
fr2/desk	0.0324	0.0313	3.40	0.0312	3.70	0.0319	1.54
fr2/xyz	0.0106	0.0104	1.89	0.0103	2.83	0.0103	2.83
fr3/long_office	0.0234	0.0226	3.42	0.0230	1.71	0.0226	3.42
fr3/sitting_xyz	0.0121	0.0117	3.31	0.0118	2.48	0.0116	4.13
fr3/sitting_static	0.0115	0.0109	5.22	0.0111	3.48	0.0107	6.96
fr3/walking_xyz	0.8439	0.8266	2.05	0.8464	−0.30	0.8115	3.84

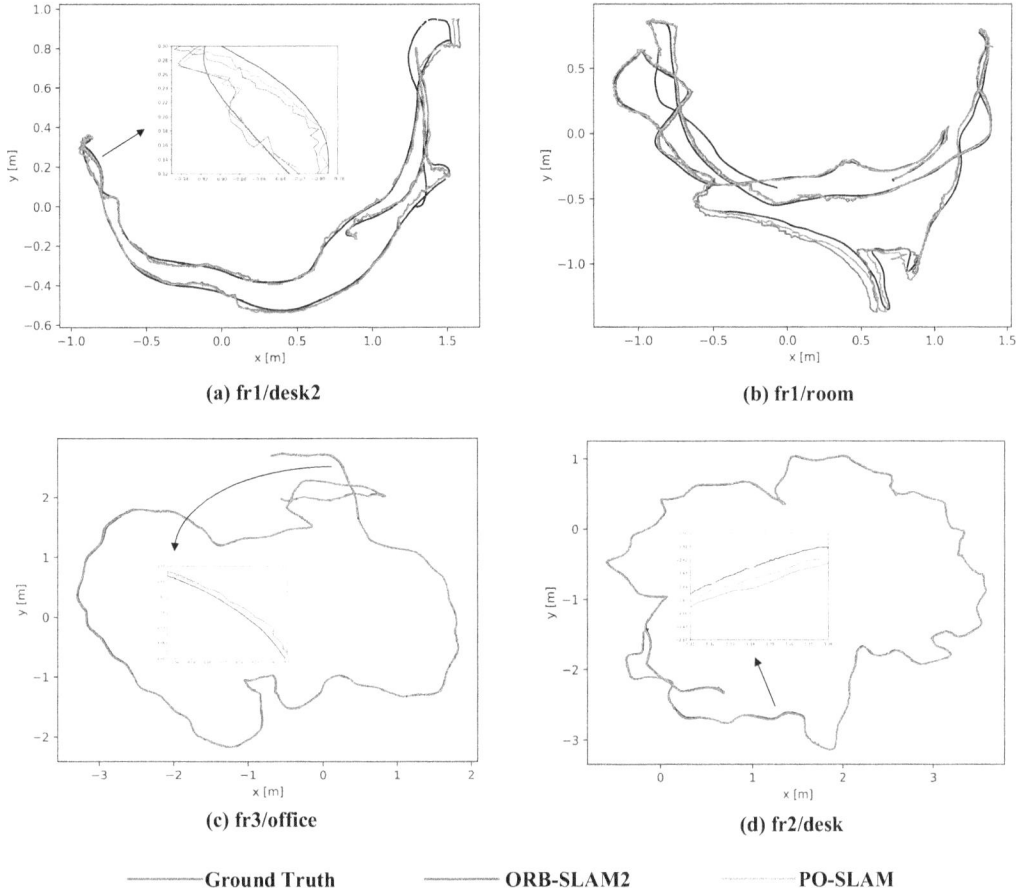

FIGURE 3.4   Comparison of trajectories estimated by our PO-SLAM, ORB-SLAM2, and ground truth on TUM RGBD dataset. (a) The trajectories on sequence fr1/desk2. (b) The trajectories on sequence fr1/room. (c) The trajectories on sequence fr3/office. (d) The trajectories on sequence fr2/desk.

proved a better adaptability to dynamic environments. Figure 3.5 illustrates a performance comparison of PO-SLAM and ORB-SLAM2 on fr3/walking_xyz dynamic sequence [22]. Clearly, ORB-SLAM2 fails to track on frames 696 and 768, while PO-SLAM is still in the SLAM mode with enough matching points with the previous frame.

The average running time per frame of PO-SLAM is demonstrated in Figure 3.6 for 10 sequences on the TUM RGBD data set. It is seen that the average time is 71.47 ms with a speed of about 14 fps, which meets the real-time requirement.

3.4 CONCLUSIONS

In this chapter, we propose a semantic visual SLAM approach combining 2D object detection and ORB feature points with additional semantic constraints for the process of BA optimization. The object segmentation approach combining object detection and the depth histogram of 2D bounding box is used to associate feature points and their corresponding

FIGURE 3.5   Comparison of our PO-SLAM and ORB-SLAM2 on fr3/walking_xyz. (a) and (b) are the results of ORB-SLAM2, and (c) and (d) describe the results of PO-SLAM.

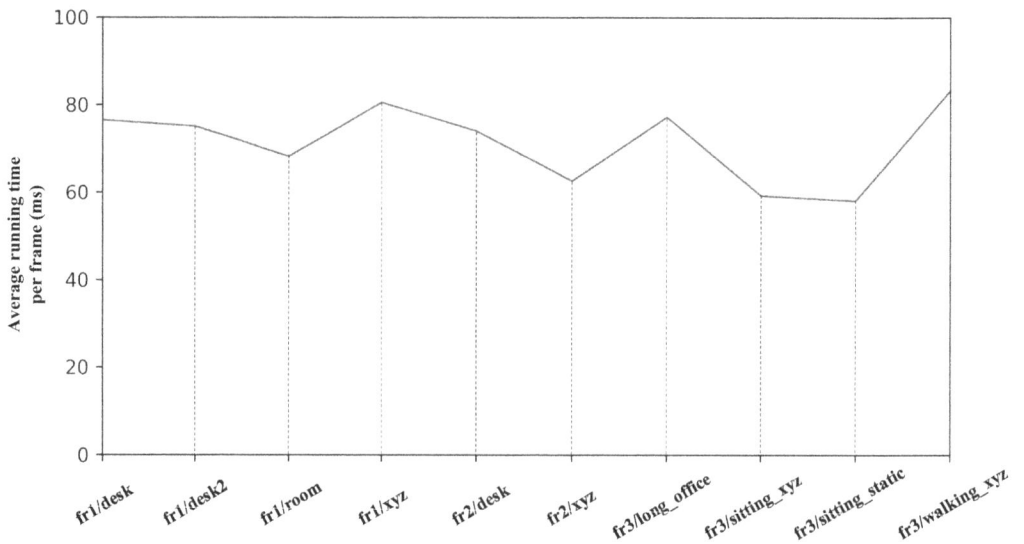

FIGURE 3.6   Average running time per frame of PO-SLAM on TUM RGBD dataset.

objects. Besides, the correlation of detected objects between two frames is also introduced. Experimental results on the TUM RGBD data set indicate that our approach can improve the accuracy and robustness compared with ORB-SLAM2.

REFERENCES

[1] Engel, J., Koltun, V., & Cremers, D. (2017). Direct sparse odometry. *IEEE Transactions on Pattern Analysis and Machine Intelligence*, 40(3), 611–625.

[2] Engel, J., Schöps, T., & Cremers, D. (2014). LSD-SLAM: Large-scale direct monocular SLAM. In *Proceedings of European Conference on Computer Vision*, Zurich, Switzerland (pp. 834–849).

[3] Klein, G., & Murray, D. (2007). Parallel tracking and mapping for small AR workspaces. In *Proceedings of IEEE and ACM International Symposium on Mixed and Augmented Reality*, Nara, Japan (pp. 225–234).

[4] Mur-Artal, R., Montiel, J. M. M., & Tardos, J. D. (2015). ORB-SLAM: A versatile and accurate monocular SLAM system. *IEEE Transactions on Robotics*, 31(5), 1147–1163.

[5] Mur-Artal, R., & Tardós, J. D. (2017). ORB-SLAM2: An open-source SLAM system for monocular, stereo, and RGB-D cameras. *IEEE Transactions on Robotics*, 33(5), 1255–1262.

[6] Zhang, L., Wei, L., Shen, P., Wei, W., Zhu, G., & Song, J. (2018). Semantic SLAM based on object detection and improved octomap. *IEEE Access*, 6, 75545–75559.

[7] Girshick, R., Donahue, J., Darrell, T., & Malik, J. (2014). Rich feature hierarchies for accurate object detection and semantic segmentation. In *Proceedings of the IEEE Conference on Computer Vision and Pattern Recognition*, Columbus, OH, USA (pp. 580–587).

[8] Girshick, R. (2015). Fast R-CNN. In *Proceedings of IEEE Conference on Computer Vision*, Santiago, Chile (pp. 1440–1448).

[9] Ren, S., He, K., Girshick, R., & Sun, J. (2016). Faster R-CNN: Towards real-time object detection with region proposal networks. *IEEE Transactions on Pattern Analysis and Machine Intelligence*, 39(6), 1137–1149.

[10] Redmon, J. (2016). You only look once: Unified, real-time object detection. In *Proceedings of the IEEE Conference on Computer Vision and Pattern Recognition*, Las Vegas, NV, USA (pp. 779–788).

[11] Redmon, J., & Farhadi, A. (2017). YOLO9000: Better, faster, stronger. In *Proceedings of the IEEE Conference on Computer Vision and Pattern Recognition*, Honolulu, HI, USA (pp. 7263–7271).

[12] Redmon, J. (2018). Yolov3: An incremental improvement. arXiv preprint arXiv:1804.02767.

[13] Liu, W., Anguelov, D., Erhan, D., Szegedy, C., Reed, S., Fu, C. Y., & Berg, A. C. (2016). SSD: Single shot multibox detector. In *Proceedings of European Conference on Computer Vision*, Amsterdam, The Netherlands (pp. 21–37).

[14] Long, J., Shelhamer, E., & Darrell, T. (2015). Fully convolutional networks for semantic segmentation. In *Proceedings of the IEEE Conference on Computer Vision and Pattern Recognition*, Boston, MA, USA (pp. 3431–3440).

[15] Badrinarayanan, V., Kendall, A., & Cipolla, R. (2017). Segnet: A deep convolutional encoder-decoder architecture for image segmentation. *IEEE Transactions on Pattern Analysis and Machine Intelligence*, 39(12), 2481–2495.

[16] He, K., Gkioxari, G., Dollár, P., & Girshick, R. (2017). Mask R-CNN. In *Proceedings of the IEEE International Conference on Computer Vision*, Venice, Italy (pp. 2961–2969).

[17] Lianos, K. N., Schonberger, J. L., Pollefeys, M., & Sattler, T. (2018). VSO: Visual semantic odometry. In *Proceedings of the European Conference on Computer Vision*, Munich, Germany (pp. 234–250).

[18] An, L., Zhang, X., Gao, H., & Liu, Y. (2017). Semantic segmentation-aided visual odometry for urban autonomous driving. *International Journal of Advanced Robotic Systems*, 14(5), 1–11.

[19] Yang, S., & Scherer, S. (2019). CubeSLAM: Monocular 3D object SLAM. *IEEE Transactions on Robotics*, 35(4), 925–938.

[20] Li, P., & Qin, T. (2018). Stereo vision-based semantic 3D object and ego-motion tracking for autonomous driving. In *Proceedings of the European Conference on Computer Vision*, Munich, Germany (pp. 646–661).

[21] Otsu, N. (1979). A threshold selection method from gray-level histograms. *IEEE Transactions on Systems, Man, and Cybernetics*, 9(1), 62–66.

[22] RGB-D SLAM Dataset and Benchmark. https://vision.in.tum.de/data/datasets/rgbd-dataset.

[23] Hartley, R., & Zisserman, A. (2003). *Multiple View Geometry in Computer Vision*. Cambridge University Press.

[24] Sturm, J., Engelhard, N., Endres, F., Burgard, W., & Cremers, D. (2012). A benchmark for the evaluation of RGB-D SLAM systems. In *Proceedings of IEEE International Conference on Intelligent Robots and Systems*, Vilamoura-Algarve, Portugal (pp. 573–580).

CHAPTER 4

Visual Relocalization from the Perspective of Scene Coordinate Regression Network

4.1 INTRODUCTION

Existing SCoRe networks use convolutional neural networks (CNNs) to predict the 3D world coordinates of each pixel in an image. However, due to the inherent geometric structure of convolutional kernels, convolution operations are inherently not invariant to geometric transformations in images caused by viewpoint changes. For a 3D point, the features of its corresponding pixels in different images from multiple viewpoints will be inevitably inconsistent, which results in that different features corresponding to the same 3D scene coordinate ground truth, and thus, the generalization is poor. This phenomenon is termed as "many-to-one" problem caused by the features that lack robustness. Besides, there exist zones with repetitive or even less texture in some scenes, which is prone to cause similar local context features at different pixel positions. In this case, the same feature appears at different locations, and it is referred to as "one-to-many" problem, where the model is possibly regressed to a mean but incorrect coordinate. Earlier CNN-based SCoRe networks usually take small local patches centered at sampled pixels as inputs [1] with the consideration of better viewpoint invariance of local patches, which alleviates the "many-to-one" problem. However, such a strategy is inefficient. Also, the local patches often cause ambiguity of patterns with weak feature discrimination, and "one-to-many" problem cannot be solved. Different from the inputs of local patches, a full RGB frame can also be sent into the SCoRe network [2–4] to capture more discriminative features, and "one-to-many" problem is effectively mitigated. Meanwhile, data augmentation including affine transformation is adopted to emulate the images in different viewpoints for the "many-to-one"

DOI: 10.1201/9781003643630-4

problem. Although the system performance is significantly improved, this emulation is still limited to model the real-world image transformations and the predicted coordinates for the same point in different images are sometimes inaccurate. The above-mentioned data augmentation is confined to the training set and may be regarded as an implicit coarse decoupling of viewpoints. How to effectively decouple features from viewpoints remains unsolved, where features of the same point in different viewpoints should be close enough.

In this chapter, a novel spatial feature transformation (SFT) network with viewpoint decoupling is proposed for a single RGB image. Inspired by the idea that each position of targets is mainly related to a local region of sources [5] where the target image is generated using a source image, the skeletons of the source and target images, our SFT aggregates the neighborhood of each source feature. It is important to note that no any prior information of target feature is provided in our method. For the neighborhood of a feature, its each element actually contributes differently to this feature, and a local attention patch is designed to represent the contribution degree of each element. For a 3D point, its corresponding features in multiple viewpoints have different local attention patches, and each patch can be generated under the guidance of global context of the image at corresponding viewpoint. Specially, the global feature is encoded as a dynamic convolutional kernel changed with input image, which is convolved with source feature map to generate parameters of local attention patches in an instance-variant manner. The obtained parameters can warp the source features to achieve viewpoint invariance. In conclusion, the main contribution of this chapter is to propose a global context-guided SFT network, which achieves the learning of the feature representation invariant to viewpoints with the constraint of maximum likelihood-derived loss. As a result, the robustness of the network to viewpoints is improved with a better solution to "many-to-one" problem. In addition, we employ the CoordConv scheme at the beginning of the SCoRe network for enhancing the position sensitivity of conventional convolutional operations [6]. The discrimination of features in repeated and texture-less zones is improved, and "one-to-many" problem is alleviated. Our scheme decouples features from their corresponding viewpoints by global context-guided SFT network to achieve a stable and accurate visual relocalization.

The rest of this chapter is structured as follows. Section 4.2 presents the related work. The proposed methodology is described in Section 4.3 in detail. Experimental results are given in Section 4.4. Section 4.5 concludes the chapter.

4.2 REVIEW OF SCORE-BASED RELOCALIZATION

During the process of scene coordinate regression, the dense 2D pixels are mapped to their corresponding 3D scene coordinates even in challenging texture-less environments. In [7–9], random forest is utilized to get the 2D–3D correspondences. Shotton et al. [7] injected RGB and depth pixel comparison features into a scene coordinate regression forest to infer the pixel-wise 3D coordinates, which is followed by a preemptive random sample consensus (RANSAC) to refine the camera pose. Rivera et al. [8] trained a set of forests to output multiple pose predictions, which were then aggregated to yield a higher pose accuracy. A joint classification regression forest [9] is designed by Brachmann et al. to model the distributions of object labels and object coordinates. With the rapid development of deep learning, the CNN-based SCoRe network has become prevalent. Brachmann et al. proposed a differentiable RANSAC method to link the coordinate CNN and score CNN for end-to-end camera pose estimation [1]. It takes local image patches as input to alleviate the influence of viewpoint change, but brings in ambiguities. Differentiable RANSAC++ (DSAC++) method [10] takes a full RGB frame as input and regresses the camera pose in an end-to-end manner with an analytic scoring strategy, where only scene coordinate network is learnable. Li et al. [3] regressed pixel-wise coordinates from a full RGB frame by a fully convolutional encoder-decoder structure. To handle "many-to-one" problem, data augmentation is imposed. Li et al. provided a hierarchical scene coordinate learning network (HSC-Net) by incorporating several classification layers into a baseline regression-only network (Reg-only) to acquire scene coordinates in a coarse-to-fine manner [2]. KFNet [4] incorporates Kalman filter in a recurrent CNN architecture to solve camera relocalization.

In the relocalization process, camera motion will inevitably cause the change of viewpoints. Under this circumstance, the same object displays different feature representations since convolutional operations are not inherently invariant to geometric transformations [11]. A common problem of existing SCoRe networks is that they lack equivariance to viewpoints. If the equivariance to viewpoints can be explicitly added, the relocalization performance is expected to be further improved. The research on equivariance has already arisen in other fields such as classification and segmentation. Jaderberg et al. presented a differentiable spatial transformer module [11], which predicts a global affine transformation matrix, and then the input features are transformed via sampling and interpolation to solve the object classification problem with part deformations. Aiming at classification and segmentation tasks, PointNet [12] comes up with a mini-network, which predicts an affine transformation matrix for input features and the transformation matrix is directly conducted matrix multiplication with input features to align them to a canonical space, so that the learned features are invariant to geometric transformations of point clouds. Note that the transformation matrix mainly focuses on the reorganization of channel components of each feature. In fact, the scheme of sampling and interpolation [11] is unsuitable for pixel-level SCoRe tasks, and the reorganization of feature channels [12] brings in a spatially-sharing transformation matrix, which is undesired for SCoRe as different spatial positions should be differentiated according to feature types such as corner and texture. Despite this, the equivariance idea from the fields of classification and segmentation provides the inspiration. In this chapter, an explicit feature transformation network with spatial discrimination is designed to achieve SCoRe with equivariance to viewpoints.

4.3 SCENE COORDINATE REGRESSION NETWORK WITH GLOBAL CONTEXT-GUIDED SPATIAL FEATURE TRANSFORMATION

A general visual relocalization method based on SCoRe network (termed as SCoRe_basic) is composed of a SCoRe network and a RANSAC-based PnP pose estimation module [3,4], where a full RGB frame is taken as input. The former is used to predict dense pixel-wise

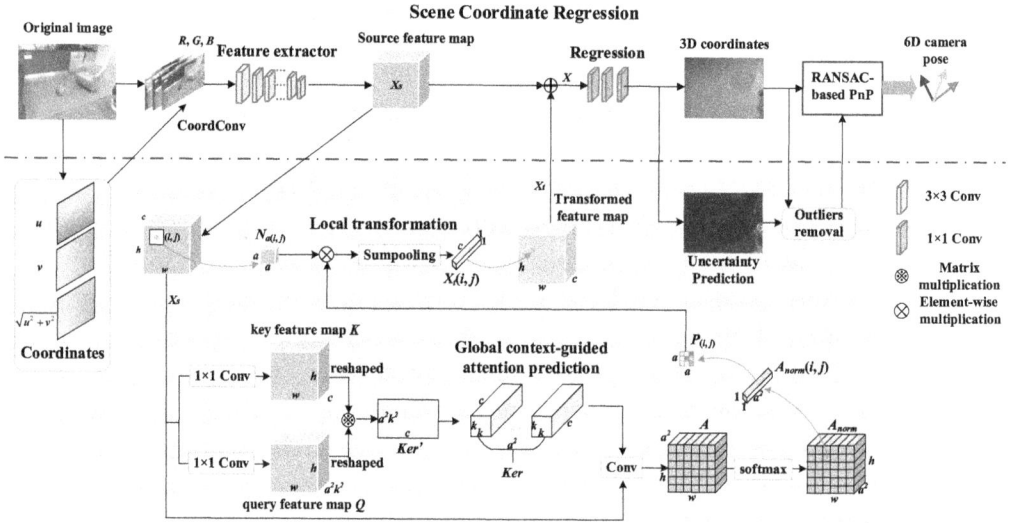

FIGURE 4.1   The pipeline of the proposed SFT-CR method for visual localization.

3D scene coordinates, which are combined with corresponding 2D pixel coordinates to solve a 6D camera pose according to the PnP algorithm. This SCoRe network adopts standard CNN architecture, which is not intrinsically invariant to viewpoint changes. In this chapter, we focus on the novel design of SCoRe network, which is still followed by PnP to achieve visual relocalization. Figure 4.1 illustrates the pipeline of the proposed spatial feature transformation-based coordinate regression (SFT-CR) for visual relocalization.

For an input RGB image, our SFT-CR first conducts feature extraction to obtain a source feature map. This feature map is transformed using SFT, and the transformed feature map, together with the source feature map, is added and sent to the regression module for 3D scene coordinates prediction. The basic frameworks of feature extractor and the regression module are similar to the first eight convolutional layers and the last three convolutional layers of the Reg-only network [2]. Due to the introduction of CoordConv scheme [6], three additional channels u, v, and $r = \sqrt{u^2 + v^2}$ are added into the first layer of our feature extractor besides the original R, G, B channels, where (u, v) is the pixel coordinate. For the last convolutional layer of the regression module, the output channels are expanded from three to four. Besides 3D coordinates, coordinate uncertainties are also outputted.

The proposed SFT network is composed of two modules: global context-guided attention prediction and point-wise local transformation. The first module predicts the parameters of corresponding local attention patch for the feature of each spatial position in the source feature map, guided by global context. To capture the global context information, the source feature map is transformed into a key feature map and a query feature map by two independent 1×1 convolutional operations. These two feature maps are reshaped to two 2D tensors, which are used to obtain Ker′. Ker′ is then further reshaped to a 4D tensor Ker, which is regarded as a global-aware dynamic convolutional kernel. Ker is convolved with the source feature map and the predicted attention matrix A is acquired, where the channels of its each spatial position record the parameters of local attention patch for this position. Different from traditional convolution, where the parameters of convolutional kernel are fixed after training, we generate a dynamic convolution kernel whose parameters are adaptively changed with the input images. The predicted attention matrix is then normalized in channel by a softmax operation so that the attention parameters for each position are scaled to [0, 1]. We denote with A_{norm} the normalized attention matrix, whose each spatial position corresponds to a local attention strip, which is then extracted and reshaped to a local attention patch. In the following second module, the neighborhood of each feature in source feature map is weighted with its local attention patch. The transformation results of all features form the transformed feature map X_t. X_t is added on the source feature map to generate the final feature map X, which is fed into the regression module for 3D coordinates and coordinate uncertainties prediction. On this basis, coordinates with large predicted uncertainties are filtered out. The filtered 2D–3D coordinate correspondences are utilized to resolve the camera pose via the RANSAC-based PnP algorithm.

It is worth noting that the dynamic convolution kernel Ker is actually an implicit representation of current viewpoint as Ker is related to the global feature. Guided by Ker, the transformed features are independent of viewpoints under the constraint of the loss function.

4.3.1 Feature Extraction

For the repeated or low-texture structure in a scene, extracting discriminative features is intractable because adjacent locations have similar local neighborhood information. To deal with this problem, CoordConv scheme in [6] is employed. Concretely, for each input image, a 3D tensor is constructed with the same spatial size for providing three channels u, v, and r. The values of each channel are scaled to [−1, 1] to keep in accordance with the range of the input. The constructed tensor is concatenated with RGB image, and the resulting tensor has the size of $8h \times 8w \times 6$ is fed into the first layer of feature extractor, which outputs the source feature map $X_s \in R^{h \times w \times c}$ whose spatial size is 1/8 of that of input image, where h, w, and c represent height, weight, and channel number of the source feature map, respectively. On the one hand, the extractor helps to differentiate different positions with similar local feature by providing additional information. On the other hand, the network parameters are increased slightly so that the increment may be neglected.

4.3.2 Global Context-Guided Attention Prediction

The purpose of this part is to generate a local attention patch with a size of $a \times a$ for each spatial position (i, j) in the source feature map with the guidance of global information, where $i \in 0,1,..,h-1$ and $j \in 0,1,..,w-1$. To take advantage of global information, we propose to encode the global feature to a dynamic convolutional kernel Ker whose spatial size is $k \times k$, which paves the foundation of point-wise local attention patches.

4.3.2.1 Global-Aware Convolutional Kernel Generation

To capture global information from the source feature map X_s in a parameter-efficient manner, we refer to the segmentation network in [13], which multiplies the matrixes corresponding to a key feature and a query feature generated by the source feature using two 1×1 convolutions. We label K and Q as the key feature map and query feature map, which enables interaction among channels of the source feature map. The size of K is the same as that of source feature map, and $K \in R^{h \times w \times c}$, whereas $Q \in R^{h \times w \times (k^2 a^2)}$ relies on the local neighboring size of features. K and Q are reshaped into $K' \in R^{hw \times c}$ and $Q' \in R^{hw \times k^2 a^2}$ by flattening the two dimensions of height and width into a dimension, respectively. After Q' is transposed, it is multiplied with K' to aggregate the features of all spatial positions in K, and a 2D tensor Ker' containing global features is acquired, where $\text{Ker}' = Q'^T K' \in R^{k^2 a^2 \times c}$. Then, Ker' is reshaped to dynamic convolutional kernel Ker with the size of $a^2 \times k \times k \times c$ by extending the dimension of $k^2 a^2$ to three dimensions of a^2, k, and k. The source feature map is convolved using Ker to generate an attention matrix A with $h \times w$ un-normalized local attention strips whose size is $1 \times 1 \times a^2$.

4.3.2.2 Local Attention Patches Prediction

By convoluting source feature map using the global-aware kernel Ker, one can obtain the predicted attention matrix $A \in R^{h \times w \times a^2}$ and normalized matrix $A_{\text{norm}} \in R^{h \times w \times a^2}$. The matrix A_{norm} consists of $h \times w$ local attention strips, each of which records the attention parameters of corresponding position and it is reshaped to a local attention patch whose size is $a \times a$. In the following, attention results are visualized. Take a frame in the Office scene of 7-Scenes dataset as an example, the local attention patches corresponding to four selected points A, B, C, and D in Figure 4.2a are depicted in Figure 4.2b, where $a = 3$. It is seen that for different pixel positions in an image, corresponding local attention patches are different.

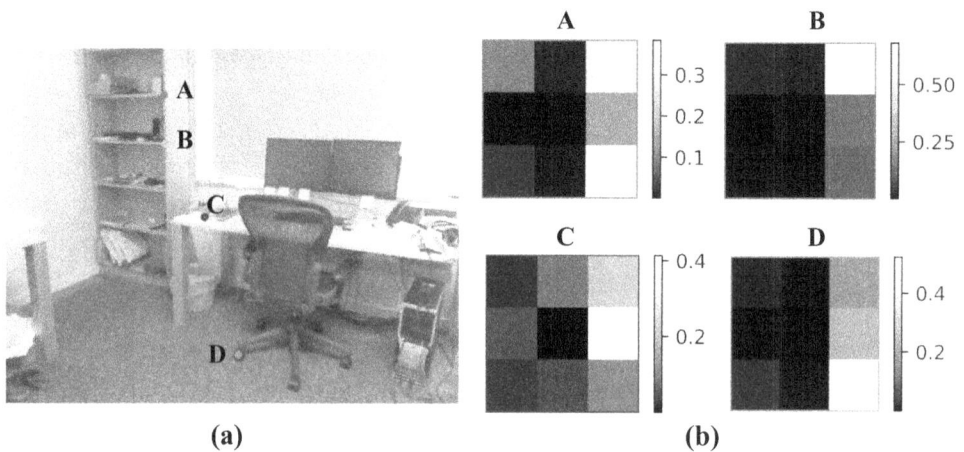

(a)   **(b)**

FIGURE 4.2 The attention visualization illustration. (a) Original RGB image with four selected positions labeled with A, B, C, and D. (b) Local attention patches corresponding to four labeled positions in (a).

Then, the local neighboring features related to each position in the source feature map are weighted using the corresponding local patch for local transformation.

4.3.3 Point-Wise Local Transformation

With the local attention patch $P_{(i,j)}$ for a position (i,j), the features $N_{a(i,j)}$ in its neighboring region can be assembled to get the transformed feature $X_t(i,j)$. The same operation is executed on each spatial position of source feature map and we obtain the transformed feature map as follows:

$$
\begin{aligned}
X_t &= \left\{ X_t(i,j) \Big|_{i=0,\ldots,h-1, j=0,\ldots,w-1} \right\} \\
&= \left\{ \left[\sum_{m=-\lfloor a/2 \rfloor}^{\lfloor a/2 \rfloor} \sum_{n=-\lfloor a/2 \rfloor}^{\lfloor a/2 \rfloor} P_{(i,j)}\left(m+\left\lfloor \frac{a}{2} \right\rfloor, n+\left\lfloor \frac{a}{2} \right\rfloor \right) \cdot X_s(i+m, j+n) \right] \Big|_{i=0,\ldots,h-1, j=0,\ldots,w-1} \right\}
\end{aligned}
\tag{4.1}
$$

where X_t is regarded as the result of local transformation. It should be noted that during the local transformation process, some information such as color and texture remains invariant. To avoid the missing of the useful information, the source feature map is combined with the transformed one by applying the residual learning strategy. Finally, the resulting feature map $X = X_s + X_t$ is sent to the regression module for coordinate prediction.

4.3.4 Maximum Likelihood-Derived Loss

For each input RGB image I, we train SFT-CR to learn the probability distribution $p(Y \mid I; \theta)$ of 3D scene coordinates rather than only the dense 3D scene coordinates, where $Y = \{Y_i \mid i = 1, 2, \ldots h \times w\}$, $Y_i = \begin{bmatrix} x_i, y_i, z_i \end{bmatrix}^T$, and θ is the network parameters. Let $f^\theta(I)$ be the network output. We formulate the training process as maximizing the logarithmic likelihood of predicted distribution:

$$
\max \log p\left(Y \mid f^\theta(I)\right) = \max \sum_{i=1}^{h \times w} \log p\left(Y_i \mid f^\theta(I)\right)
\tag{4.2}
$$

Inspired by [14], three components of Y_i, namely, x_i, y_i, and z_i are independent of each other, and each component follows the univariate Laplace distribution:

$$
L(x \mid a, b) = \frac{1}{2b} e^{-\frac{|x-a|}{b}}
\tag{4.3}
$$

where a is the expectation of Laplace distribution. The variance of Laplace distribution is $2b^2$, and b can be seen as the reflection of uncertainty of predicted 3D coordinate.

By minimizing the negative logarithmic likelihood, the loss function L_r is given by

$$L_r = -\sum_{i=1}^{h \times w} \log p\left(Y_i \mid f^\theta(I)\right)$$

$$= -\sum_{i=1}^{h \times w} \log\left[p\left(x_i \mid f^\theta(I)\right) p\left(y_i \mid f^\theta(I)\right) p\left(z_i \mid f^\theta(I)\right) \right]$$

$$= -\sum_{i=1}^{h \times w} \left[\log p\left(x_i \mid f^\theta(I)\right) + \log p\left(y_i \mid f^\theta(I)\right) + \log p\left(z_i \mid f^\theta(I)\right) \right]$$

$$= \sum_{i=1}^{h \times w} \left[\frac{\left|x_i - a_i^x\right|}{b_i^x} + \frac{\left|y_i - a_i^y\right|}{b_i^y} + \frac{\left|z_i - a_i^z\right|}{b_i^z} + \log b_i^x + \log b_i^y + \log b_i^z + 3\log 2 \right]$$

(4.4)

Compared with the uncertainties of three components of each predicted 3D coordinate, we are more concerned with the uncertainty of the whole coordinate. For simplicity, x_i, y_i, and z_i for each point i have the same parameter b_i, namely, $b_i^x = b_i^y = b_i^z = b_i$. b_i and $a_i = \left[a_i^x, a_i^y, a_i^z \right]^T$ are outputted by the network.

Neglecting the constant term of (4.4), we have

$$L_r = \sum_{i=1}^{h \times w} \left[\frac{\left|x_i - a_i^x\right| + \left|y_i - a_i^y\right| + \left|z_i - a_i^z\right|}{b_i} + 3\log b_i \right]$$

$$\leq \sqrt{3} \sum_{i=1}^{h \times w} \left[\frac{\sqrt{\left(x_i - a_i^x\right)^2 + \left(y_i - a_i^y\right)^2 + \left(z_i - a_i^z\right)^2}}{b_i} + \sqrt{3}\log b_i \right]$$

(4.5)

Alternatively, we consider the upper limitation of L_r to represent the final maximum likelihood-derived loss L_{fr}:

$$L_{fr} = \sqrt{3} \sum_{i=1}^{h \times w} M_i \left(\frac{\sqrt{\left(x_i - a_i^x\right)^2 + \left(y_i - a_i^y\right)^2 + \left(z_i - a_i^z\right)^2}}{b_i} + \sqrt{3}\log b_i \right)$$

(4.6)

where M_i indicates whether the 3D ground truth coordinate of point i exists. If it exists, M_i is 1; otherwise, it is 0. To avoid the occurrence of gradient exploding, the network outputs $\log b_i$ rather than b_i.

Different from traditional Euclidean distance loss that treats each error between prediction and ground truth equally, we can adaptively reweight the Euclidean distance through b_i. When the prediction of coordinate is far away from its ground truth, minimizing L_{fr} will force b_i to get larger. As a result, the influence of outliers is relieved and the network more focuses on the predictions with smaller errors.

4.3.5 Pose Estimation

In this part, we employ RANSAC-based PnP algorithm [2] to figure out the camera pose $\left[R^*, t^*\right]$ using the predicted dense 2D–3D correspondences $\{p_i \mid i=1,2,\ldots,h \times w\}$ and $\{Y_i \mid i=1,2,\ldots,h \times w\}$. Firstly, the 2D–3D point pairs whose uncertainty is larger than a threshold σ_u are filtered out. Then K_p point sets are constructed by random sampling, where each set contains 4 pairs. Correspondingly, K_p pose hypotheses are computed by PnP algorithm. Each hypothesis $\left[R, t\right]$ is scored according to the soft inlier counting in [10], where R and t are estimated rotation matrix and translation vector, respectively. The hypothesis with the maximal score is further refined and optimal pose $[R^*, t^*]$ is then obtained.

4.4 EXPERIMENTS

In this section, the experiments are conducted to demonstrate the proposed SFT-CR method as well as the comparison with existing state-of-the-art works on two public data-sets: 7-scenes and 12-scenes. 7-Scenes dataset [7] is an indoor RGB-D dataset provided by Microsoft, which consists of seven different indoor scenes with various challenges such as motion blur, viewpoint variance, low-texture surfaces, and repeated structures. All scenes are recorded using a handheld Kinect camera at a resolution of 640×480. Every scene is composed of several sequences, which are divided into training set and testing set. Besides RGB and depth images, the ground truth of camera poses and dense 3D models are provided. 12-Scenes dataset [15] contains RGB-D data of 12 different scenes. The RGB frames are recorded with an iPad color camera with a resolution of $1,296 \times 968$, while the depth frames are taken from a depth sensor with 640×480 resolution.

With RGB, depth image, and ground truth of pose at each frame, we acquire the ground truth of pixel-wise 3D scene coordinates. Notice that the depth images are dismissed and a RGB-based relocalization is achieved at the testing stage. For each training sample, data augmentation is executed according to [2] for sample diversity. We train our model from scratch with 200–300k iterations for different scenes. The Adam optimizer is used with $\beta_1 = 0.9$, $\beta_2 = 0.999$, and $\epsilon = 10^{-8}$.

4.4.1 Ablation Studies

Firstly, the values of hyper-parameters k and a are determined, where k is the spatial size of dynamic convolutional kernel and a is the size of local attention patch. The results with different k and a on the Office scene of 7-Scenes dataset are shown in Table 4.1. The median pose error (Med. Err.) consists of the median translation and rotation errors of estimated poses relative to ground truth, and the 5 cm–5° accuracy (Acc.) reflects the proportion of estimated poses whose translation and rotation errors are under 5 cm and 5°. Combining the metrics of Acc. and Med. Err., we select $k = a = 3$ in this chapter.

To testify the performance of our proposed SFT-CR method, its five variants SFT-CR-I, SFT-CR-II, SFT-CR-III, SFT-CR-IV, SFT-CR-V, and SCoRe_basic are involved according to whether CoordConv, maximum likelihood-derived loss, convolutional transformation, as well as global guidance and local transformation of SFT are considered. The convolutional transformation composed of three stacked standard convolutions (SFT-CR-III) is used to replace SFT network with similar model size. This is beneficial to verify our proposed SFT network.

TABLE 4.1 The Comparison of Different k and a on the Office Scene of 7-Scenes Dataset in Terms of Median Pose Error (m, °) and 5 cm–5° Accuracy (%)

k	a	Acc.	Med. Err.	k	a	Acc.	Med. Err.
1	1	86.05%	0.026 m, 0.76°	1	3	89.33%	0.025 m, 0.69°
3	1	86.28%	0.027 m, 0.74°	3	3	90.63%	0.024 m, 0.66°
5	1	86.98%	0.025 m, 0.73°	5	3	90.38%	0.025 m, 0.69°
7	1	86.52%	0.026 m, 0.71°	7	3	89.22%	0.025 m, 0.67°
1	5	89.13%	0.024 m, 0.67°	1	7	87.42%	0.025 m, 0.74°
3	5	89.98%	0.025 m, 0.71°	3	7	87.95%	0.025 m, 0.69°
5	5	90.48%	0.023 m, 0.67°	5	7	87.02%	0.024 m, 0.70°
7	5	88.75%	0.024 m, 0.71°	7	7	89.85%	0.024 m, 0.70°

TABLE 4.2 The Comparison of Different Variants of Our SFT-CR Method Averaged Over All Scenes of 7-Scenes Dataset in Terms of Median Pose Error (m, °) and 5 cm–5° Accuracy (%)

Method	Coord Conv	Maximum Likelihood Loss	Convolutional Transformation	Global Guidance	Local Transformation	Model Size	Acc.	Med. Err.
SCoRe_basic	—	—	—	—	—	104 M	78.8%	0.031 m, 0.99°
SFT-CR-I	—	✓	—	—	—	104 M	81.8%	0.028 m, 0.94°
SFT-CR-II	✓	✓	—	—	—	104 M	84.0%	0.028 m, 0.89°
SFT-CR-III	✓	✓	✓	—	—	105 M	80.3%	0.029 m, 0.95°
SFT-CR-IV	✓	✓	—	—	✓	106 M	83.0%	0.028 m, 0.92°
SFT-CR-V	✓	✓	—	✓	—	114 M	82.6%	0.027 m, 0.88°
SFT-CR	✓	✓	—	✓	✓	105 M	86.1%	0.026 m, 0.85°

Besides, the SFT-CR-IV and SFT-CR-V methods refer to the solutions of SFT without global guidance and local transformation, respectively, where the former obtains the attention matrix by three standard convolutions with 3×3 kernels operated on the source feature map, whereas the latter removes local transformation and the number of output channels of dynamic convolution is in coincidence with the channel number of source feature map.

Table 4.2 presents the comparison results averaged over all scenes of the 7-Scenes dataset in terms of median pose error and 5 cm–5° accuracy. Comparing the results of SCoRe_basic and SFT-CR-I, the average relocalization accuracy is improved with the help of maximum likelihood-derived loss. This factor is combined with CoordConv to achieve a better performance, as seen in the results of SFT-CR-II. Also, the advantage of the proposed SFT network can be supported by the comparison with SFT-CR-III, which replaces SFT using convolutional transformation. Finally, the results of SFT-CR-IV, SFT-CR-V, and SFT-CR indicate that the combination of global context-guided attention prediction and local transformation is more effective.

4.4.2 Relocalization Accuracy

We evaluate the relocalization accuracy of our SFT-CR quantitatively. Table 4.3 gives the median pose errors of our method and existing methods including Active Search [16], InLoc [17], PoseNet [18], SCRF [7], Auto-context [9], DSAC++ [10], Reg-only [2], HSC-Net [2],

TABLE 4.3 Comparison of Different Methods on 7-Scenes Dataset in Terms of Median Pose Error (m, °)

7-Scenes	Active Search	Inloc	PoseNet	SCRF	Auto-Context	DSAC++	Reg-Only	HSC-Net	SCoordNet	KFNet	SFT-CR
Chess	0.04, 1.96	0.03, 1.05	0.32, 4.06	0.03, 0.66	**0.015**, 1.3	0.02, **0.5**	0.021, 0.70	0.021, 0.68	0.019, 0.63	0.018, 0.65	0.021, 0.70
Fire	0.03, 1.53	0.03, 1.07	0.47, 7.33	0.05, 1.50	0.030, 1.4	**0.02**, 0.9	0.024, 0.91	0.022, 0.87	0.023, 0.91	0.023, 0.90	**0.020**, **0.78**
Heads	0.02, 1.45	0.02, 1.16	0.29, 6.00	0.06, 5.50	0.059, 3.4	**0.01**, **0.8**	0.012, 0.82	0.012, 0.86	0.018, 1.26	0.014, 0.82	0.011, 0.81
Office	0.09, 3.61	0.03, 1.05	0.48, 3.84	0.04, 0.78	0.047, 1.7	0.03, 0.7	0.031, 0.92	0.027, 0.79	0.026, 0.73	0.025, 0.69	**0.024**, **0.66**
Pumpkin	0.08, 3.10	0.05, 1.55	0.47, 4.21	0.04, **0.68**	0.043, 2.1	0.04, 1.1	0.043, 1.14	0.040, 1.02	0.039, 1.09	0.037, 1.02	**0.034**, 0.98
Redkitchen	0.07, 3.37	0.04, 1.31	0.59, 4.32	0.04, **0.76**	0.058, 2.2	0.04, 1.1	0.045, 1.40	0.040, 1.18	0.039, 1.18	0.038, 1.16	**0.034**, 1.06
Stairs	**0.03**, 2.22	0.09, 2.47	0.47, 6.93	0.32, 1.32	0.174, 7.0	0.09, 2.6	0.038, 1.03	0.031, **0.82**	0.037, 1.06	0.033, 0.94	0.035, 0.97
Average	0.051, 2.46	0.041, 1.38	0.441, 5.24	0.083, 1.60	0.061, 2.73	0.036, 1.10	0.031, 0.99	0.028, 0.89	0.029, 0.98	0.027, 0.88	**0.026**, **0.85**

SCoordNet [4], and KFNet [4] on 7-Scenes dataset. The first two methods belong to the feature matching-based type; the third one corresponds to direct pose regression; the fourth and fifth are classified as SCoRe with random forest, whereas the last five methods are all CNN-based SCoRe solutions. It is noted that SCoordNet refers to the single-frame version of KFNet. As shown in Table 4.3, our SFT-CR achieves the lowest average median pose error on 7-Scenes dataset, which manifests the effectiveness of our method. Compared with HSC-Net, our method omits the hierarchical discrete location labeling at the training phase.

Table 4.4 illustrates the comparison of SFT-CR with Reg-only and HSC-Net in terms of median pose error on 12-Scenes dataset. Clearly, all three methods achieve better performances, which may be because the training and test trajectories on 12-Scenes are close without significant viewpoint changes between training and test frames [2]. In Table 4.4, the average relocalization accuracy of our SFT-CR is slightly higher than that of the other two methods.

Table 4.5 presents the number of frames with accuracy improvement of our method over Reg-only and HSC-Net on 7-Scenes and 12-Scenes datasets. N_{ft1} and N_{fr1} are labeled as numbers of frames for our method whose translation error and rotation error are better than Reg-only, respectively. We denote with N_{ft2} and N_{fr2} the numbers of frames for our method whose translation error and rotation error are better than HSC-Net, respectively. It is seen that our method possesses a larger proportion of frames with higher pose accuracy than Reg-only and HSC-Net. Combining the results of Tables 4.3 and 4.4, the proposed method with global context-guided attention prediction and local transformation is considered as an effective one.

TABLE 4.4   Comparison of Different Methods on 12-Scenes Dataset in Terms of Median Pose Error (m, °)

12-Scenes	Reg-only	HSC-Net	SFT-CR
Kitchen-1	0.008, **0.4**	0.008, **0.4**	**0.007, 0.4**
Living-1	0.011, **0.4**	0.011, **0.4**	**0.010, 0.4**
Bed	0.013, 0.6	**0.009, 0.4**	0.011, **0.4**
Kitchen-2	0.008, 0.4	**0.007, 0.3**	**0.007**, 0.4
Living-2	0.014, 0.6	**0.010, 0.4**	0.011, **0.4**
Luke	0.020, 0.9	0.012, **0.5**	**0.011, 0.5**
Gates 362	0.011, 0.5	0.010, **0.4**	**0.009, 0.4**
Gates 381	0.016, 0.7	**0.012**, 0.6	0.013, **0.5**
Lounge	0.015, 0.5	0.014, 0.5	**0.012, 0.4**
Manolis	0.014, 0.7	0.011, **0.5**	**0.010, 0.5**
Floor5a	0.016, 0.7	**0.012, 0.5**	**0.012, 0.5**
Floor5b	0.019, 0.6	0.015, **0.5**	**0.014, 0.5**
Average	0.014, 0.6	**0.011**, 0.5	**0.011, 0.4**

TABLE 4.5   Numbers of Frames with Improved Accuracy of Our Method Compared with Reg-Only and HSC-Net on 7-Scenes and 12-Scenes

Dataset	Number of All Test Frames	Nft1	Nfr1	Nft2	Nfr2
7-Scenes	17,000	12,091	12,316	10,733	10,604
12-Scenes	5,702	3,890	3,832	3,069	3,011

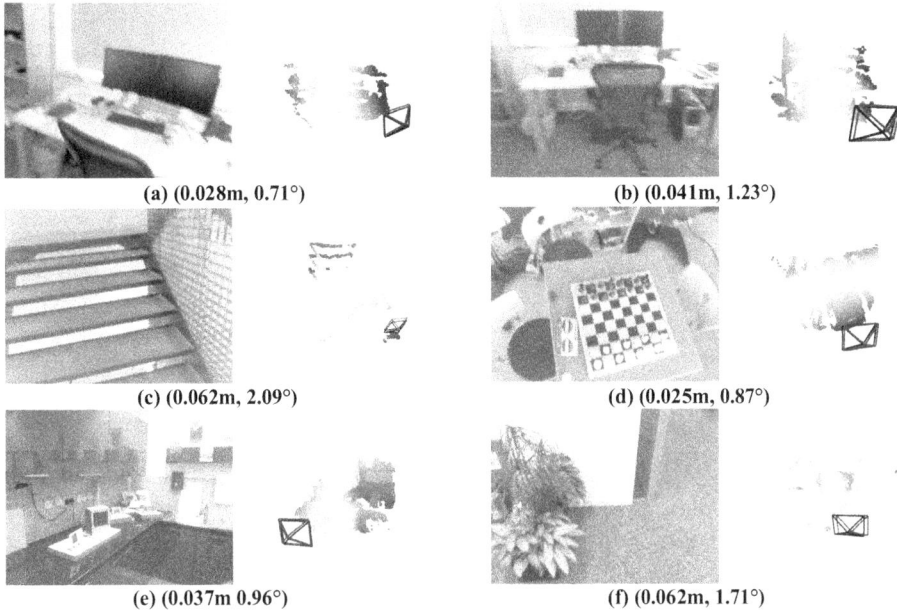

FIGURE 4.3   The translation error (m) and rotation error (°) for six images with different challenging conditions. (a, b) Motion blur. (c, d) Repeated structures. (e) Specularity. (f) Low-texture surface.

In the following, we select six images with different challenging conditions including motion blur, repeated structures, specularity, and low-texture surface from the test set of 7-Scene dataset, and the translation error and rotation error are provided in Figure 4.3. The original RGB image, the predicated scene coordinates, the estimated camera pose, and the ground truth pose for each selected image are also shown. The results show that the camera poses estimated by the proposed method are acceptable.

Besides, artificial changes in blur and brightness are also imposed on two images from the test set of 7-Scenes dataset. Figure 4.4 illustrates the translation error and rotation error. For the original RGB image illustrated in Figure 4.4a, Gaussian blur with a kernel size of 31×31 pixels and standard deviations σ_x and σ_y is applied. We obtain Figure 4.4a1 and a2 by setting $\sigma_x = \sigma_y = 3$ and $\sigma_x = \sigma_y = 5$, respectively. In addition, we change the brightness of the original RGB image shown in Figure 4.4b by adding an increment B to all pixels. Figure 4.4b1 and b2 is generated when B is set to 50 and 100, respectively. In spite of perturbations, the proposed method can still output feasible results.

4.4.3 Scene Coordinate Accuracy

Our scene coordinate regression network outputs pixel-wise scene coordinates for each image. To quantitatively analyze scene coordinate errors, the cumulative distribution function (CDF) is selected, where CDF reflects the proportion of pixels whose coordinate errors between predictions and ground truth are within a given threshold. Figure 4.5 presents the CDF results of SFT-CR, Reg-only, and HSC-Net on 2 scenes from 7-Scenes and

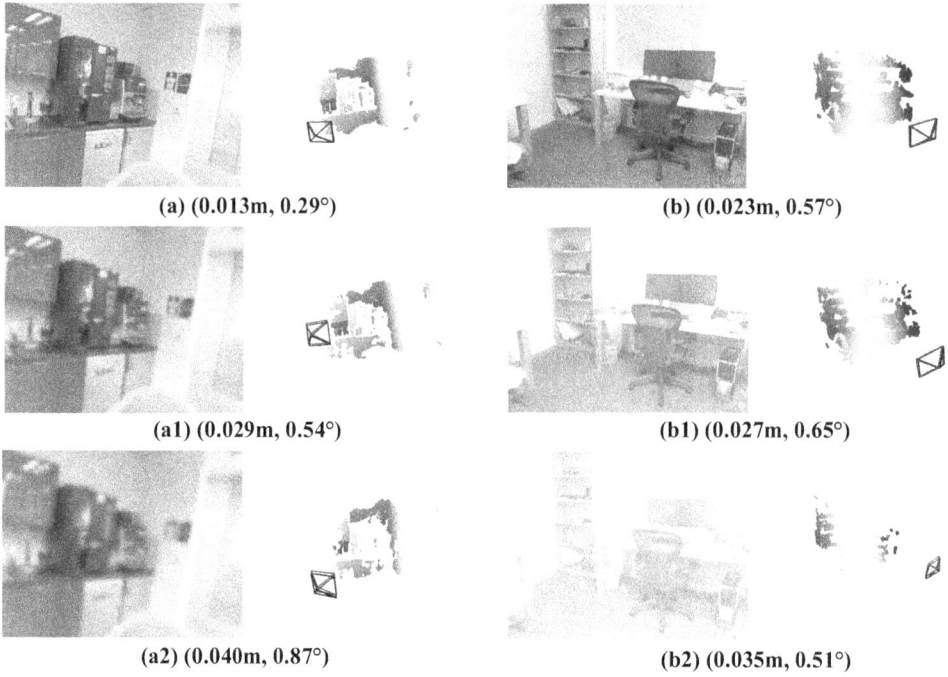

(a) (0.013m, 0.29°) (b) (0.023m, 0.57°)

(a1) (0.029m, 0.54°) (b1) (0.027m, 0.65°)

(a2) (0.040m, 0.87°) (b2) (0.035m, 0.51°)

FIGURE 4.4 The translation error (m) and rotation error (°) for the images. (a, a1, a2) represent the images with varying blur. (b, b1, b2) represent the images with varying brightness.

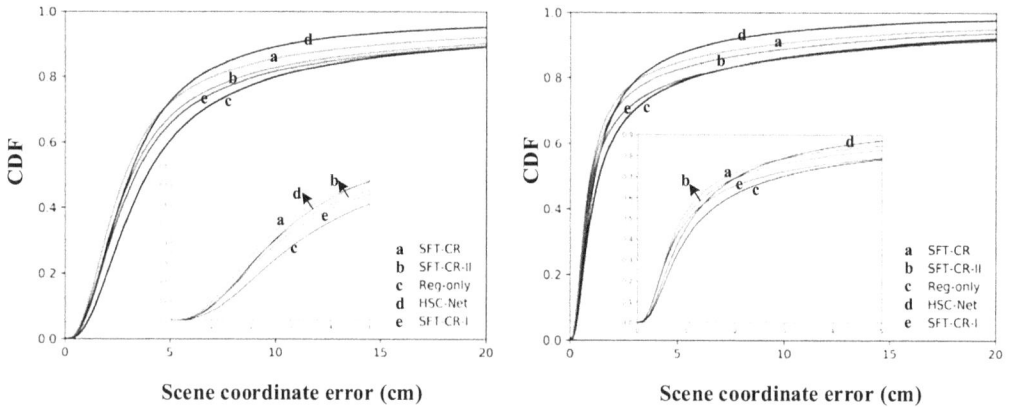

FIGURE 4.5 Cumulative distribution functions (CDF) of scene coordinate errors for different solutions on two scenes. (a) Office scene and (b) Kitchen-1 scene.

12-Scenes datasets. Also, SFT-CR-I and SFT-CR-II are considered to further declare the proposed spatial feature transformation network. Obviously, the cumulative distribution curve of our SFT-CR method is always above that of Reg-only. As Zhou et al. pointed out [4], precise scene coordinates, especially those with errors smaller than 2 cm, are crucial

TABLE 4.6   Average Relocalization Time of Different Methods on
7-Scenes Dataset

Methods	KFNet	Reg-Only	HSC-Net	SFT-CR
Averaged time per frame	>155 ms	153 ms	165 ms	124 ms

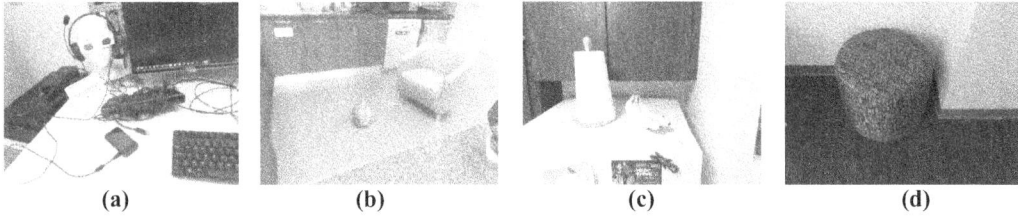

| (a) | (b) | (c) | (d) |

FIGURE 4.6   Four selective frames from different scenes in 7-Scenes and 12-Scenes datasets. Frames (a, b) are from Heads and Pumpkin scenes, and frames (c, d) belong to Kitchen-1 and Floor5a scenes, respectively.

to the accuracy of pose estimation. Therefore, we pay more attention to the proportion of pixels corresponding to smaller scene coordinate errors. From the CDF curves of HSC-Net and SFT-CR in Figure 4.5, we can see that our method occupies a larger proportion of coordinate predictions within small scene coordinate error. This phenomenon indicates our predicted coordinates can acquire better pose accuracy, which can also be seen in Tables 4.3 and 4.4. Besides, the CDF result of SFT-CR is consistently superior to that of SFT-CR-I and SFT-CR-II.

4.4.4 Efficiency

In this part, we further investigate the efficiency of the proposed SFT-CR method. With the platform of NVIDIA GeForce GTX 1080 Ti GPU and Intel Xeon E5–2650 v4 CPU, the average relocalization time per frame of different methods on 7-Scenes dataset is reported in Table 4.6. It is seen that our SFT-CR executes the visual relocalization at a faster speed. The proposed method is also run only on the CPU, and the average processing time per frame on 7-Scenes dataset is 903 ms, which is acceptable for the relocalization application that is mainly activated when the tracking fails.

4.4.5 Scene Recognition Test Based on Coordinate Uncertainty

The uncertainty of coordinate prediction reflects that we can trust the predicted coordinate with what confidence. A potential application of uncertainty is scene recognition, where the scene class is recognized for a test image. Considering the 19 scene classes of 7-Scenes and 12-Scenes datasets, the mean uncertainty of each selected frame (see Figure 4.6) on our trained SFT-CR model of each scene is presented in Table 4.7. The scene class corresponding to the minimal mean uncertainty value is selected, and the recognized result of each test frame is consistent with its real scene.

TABLE 4.7 The Mean Uncertainties of Four Selected Frames from Figure 4.6 on Each of the Models Corresponding to 19-Scenes

	Chess	Fire	Heads	Office	Pumpkin	Redkitchen	Stairs	Kitchen-1	Living-1	Bed	Kitchen-2	Living-2	Luke	Gates 362	Gates 381	Lounge	Manolis	Floor5a	Floor5b
(a)	0.206	0.166	0.020	0.253	0.344	0.410	0.409	0.216	0.531	0.171	0.310	0.281	0.188	0.311	0.453	0.395	0.206	0.220	0.266
(b)	0.235	0.193	0.123	0.315	0.056	0.419	0.357	0.331	0.749	0.172	0.247	0.285	0.261	0.285	0.296	0.464	0.319	0.606	0.361
(c)	0.213	0.229	0.187	0.343	0.323	0.398	0.595	0.017	0.569	0.271	0.406	0.446	0.236	0.482	0.492	0.621	0.342	0.424	0.348
(d)	0.169	0.124	0.128	0.232	0.265	0.311	0.425	0.217	0.397	0.180	0.168	0.226	0.367	0.342	0.266	0.321	0.352	0.032	0.188

4.5 CONCLUSION

In this chapter, we promote the SCoRe network-based visual relocalization by improving the robustness and discrimination of intermediate CNN features. An SFT network is proposed, which uses a dynamic convolutional kernel containing global information to guide the local transformation on source feature map. Therefore, the invariance of feature to viewpoints is achieved under the constraint of the loss function. CoordConv is employed to improve the discrimination of features. Experimental results on 7-Scenes and 12-Scenes datasets verify the effectiveness of the proposed method. In our future work, we shall focus on more indoor and outdoor datasets with ground truth acquisition based on motion-tracking to further verify the proposed method. In addition, multi-modal distribution [19] is considered a more reliable prediction of uncertainty.

REFERENCES

[1] Brachmann, E., Krull, A., Nowozin, S., Shotton, J., Michel, F., Gumhold, S., & Rother, C. (2017). DSAC-differentiable RANSAC for camera localization. In *Proceedings of the IEEE Conference on Computer Vision and Pattern Recognition*, Honolulu, HI, USA (pp. 6684–6692).

[2] Li, X., Wang, S., Zhao, Y., Verbeek, J., & Kannala, J. (2020). Hierarchical scene coordinate classification and regression for visual localization. In *Proceedings of the IEEE Conference on Computer Vision and Pattern Recognition*, Seattle, WA, USA (pp. 11983–11992).

[3] Li, X., Ylioinas, J., & Kannala, J. (2018). Full-frame scene coordinate regression for image-based localization. arXiv:1802.03237.

[4] Zhou, L., Luo, Z., Shen, T., Zhang, J., Zhen, M., Yao, Y., Fang, T., & Quan, L. (2020). KFNet: Learning temporal camera relocalization using Kalman filtering. In *Proceedings of the IEEE Conference on Computer Vision and Pattern Recognition*, Seattle, WA, USA (pp. 4919–4928).

[5] Ren, Y., Yu, X., Chen, J., Li, T. H., & Li, G. (2020). Deep image spatial transformation for person image generation. In *Proceedings of the IEEE Conference on Computer Vision and Pattern Recognition*, Seattle, WA, USA (pp. 7690–7699).

[6] Liu, R., Lehman, J., Molino, P., Petroski Such, F., Frank, E., Sergeev, A., & Yosinski, J. (2018). An intriguing failing of convolutional neural networks and the CoordConv solution. In *Advances in Neural Information Processing Systems,* Montréal, Canada (pp. 9628–9639).

[7] Shotton, J., Glocker, B., Zach, C., Izadi, S., Criminisi, A., & Fitzgibbon, A. (2013). Scene coordinate regression forests for camera relocalization in RGB-D images. In *Proceedings of the IEEE Conference on Computer Vision and Pattern Recognition*, Portland, OR, USA (pp. 2930–2937).

[8] Guzman-Rivera, A., Kohli, P., Glocker, B., Shotton, J., Sharp, T., Fitzgibbon, A., & Izadi, S. (2014). Multi-output learning for camera relocalization. In *Proceedings of the IEEE Conference on Computer Vision and Pattern Recognition*, Columbus, OH, USA (pp. 1114–1121).

[9] Brachmann, E., Michel, F., Krull, A., Yang, M. Y., & Gumhold, S. (2016). Uncertainty-driven 6D pose estimation of objects and scenes from a single RGB image. In *Proceedings of the IEEE Conference on Computer Vision and Pattern Recognition*, Las Vegas, NV, USA (pp. 3364–3372).

[10] Brachmann, E., & Rother, C. (2018). Learning less is more-6D camera localization via 3D surface regression. In *Proceedings of the IEEE Conference on Computer Vision and Pattern Recognition*, Salt Lake City, UT, USA (pp. 4654–4662).

[11] Jaderberg, M., Simonyan, K., & Zisserman, A. (2015). Spatial transformer networks. In *Advances in Neural Information Processing Systems,* Cambridge, MA, UK (pp. 2017–2025).

[12] Qi, C. R., Su, H., Mo, K., & Guibas, L. J. (2017). Pointnet: Deep learning on point sets for 3D classification and segmentation. In *Proceedings of the IEEE Conference on Computer Vision and Pattern Recognition*, Honolulu, HI, USA (pp. 652–660).

[13] Liu, J., He, J., Qiao, Y., Ren, J. S., & Li, H. (2020). Learning to predict context-adaptive convolution for semantic segmentation. In *Proceedings of the European Conference on Computer Vision*, Glasgow, UK (pp. 769–786).

[14] Ilg, E., Cicek, O., Galesso, S., Klein, A., Makansi, O., Hutter, F., & Brox, T. (2018). Uncertainty estimates and multi-hypotheses networks for optical flow. In *Proceedings of the European Conference on Computer Vision*, Munich, Germany (pp. 652–667).

[15] Valentin, J., Dai, A., Nießner, M., Kohli, P., Torr, P., Izadi, S., & Keskin, C. (2016). Learning to navigate the energy landscape. In *Proceedings of the International Conference on 3D Vision*, Stanford, CA, USA (pp. 323–332).

[16] Sattler, T., Leibe, B., & Kobbelt, L. (2016). Efficient & effective prioritized matching for large-scale image-based localization. *IEEE Transactions on Pattern Analysis and Machine Intelligence*, 39(9), 1744–1756.

[17] Taira, H., Okutomi, M., Sattler, T., Cimpoi, M., Pollefeys, M., Sivic, J., Pajdla, T., & Torii, A. (2018). InLoc: Indoor visual localization with dense matching and view synthesis. In *Proceedings of the IEEE Conference on Computer Vision and Pattern Recognition*, Salt Lake City, UT, USA (pp. 7199–7209).

[18] Kendall, A., Grimes, M., & Cipolla, R. (2015). Posenet: A convolutional network for real-time 6-DOF camera relocalization. In *Proceedings of the IEEE International Conference on Computer Vision*, Santiago, Chile (pp. 2938–2946).

[19] Truong, P., Danelljan, M., Van Gool, L., & Timofte, R. (2021). Learning accurate dense correspondences and when to trust them. In *Proceedings of the IEEE Conference on Computer Vision and Pattern Recognition*, Nashville, TN, USA (pp. 5714–5724).

CHAPTER 5

Visual Relocalization from the Perspective of Place Recognition

5.1 INTRODUCTION

Visual place recognition is generally formulated as a retrieval-based localization problem with an environment map including a set of geo-tagged reference images. The geographic location of a query image can usually be approximated with that of the most similar reference image. Such a process is similar to human reasoning, where human localizes themselves by recognizing the salient landmarks previously seen. However, place recognition suffers from severe challenges because the observation images at the same place vary dramatically due to different viewpoints, illuminations, seasons, weather, and even dynamic objects, while the images taken at different places may have a similar appearance. The robustness of these variations is desired.

For place recognition, the key part is to learn a discriminative global feature representation to describe an image, which means that the images taken at the same place hold similar image features, whereas the features to describe those images taken at different places are not similar [1]. The traditional solutions first extract local feature descriptors such as scale-invariant feature transforms (SIFT) [2] and then congregate them by bag of words (BoW) [3] or vector of locally aggregated descriptors (VLAD) [4] to obtain a holistic descriptor for the whole image. Finally, the place of query image can be recognized by computing the distance of features between the query image and each reference image. Such handcrafted features are vulnerable with weak adaptability to complex environmental variations [5]. Due to the extraordinary capability of feature representation on various computer vision tasks [6], convolutional neural network (CNN) has become prevailing, where discriminative image features to characterize the places are attained. The mainstream CNN-based place recognition is treated as a ranking task, which can be solved with the help of deep metric learning [7–13]. Although these methods made some improvements compared to the previous ones, there still exist drawbacks. On the one hand, not every location in an image contributes equally to place recognition since humans can recognize a place only through certain salient landmarks. Furthermore, some regions are occupied by some dynamic or undistinguishable objects, which could cause interference and should be neglected. Existing attention mechanism-based methods [14–16] emphasize the important regions by capturing more flexible context information. A possible problem is that they pay less attention on the robustness to image appearance variations caused by various environmental conditions. More importantly, in the common triplet ranking task, each positive pair of a query image is required to hold smaller feature distances than corresponding negative pairs. This makes the triplet ranking task more concerned with the similarity to query image in appearance [17]. If two images are taken from the same place with severe environmental change, their similarity in appearance becomes small, which will cause a large feature distance. It is still challenging to recognize such a true positive pair by only training with the ranking task. Such a case is unfavorable to the generalization performance of the model for unseen test images [17,18].

To solve the aforementioned problems, this chapter proposes a multi-task learning method with attentive feature aggregation for visual place recognition. The main contributions of this chapter are twofold. Firstly, a multi-task learning method is proposed, which takes full advantage of classification and ranking tasks to train the network jointly. Particularly, a binary classification task is introduced, where two classes are constructed: positive and negative ones. The positive class covers all query-positive pairs from the triplet ranking task, and all query-negative pairs constitute the negative class. Then, a binary classification network with a binary classification loss is designed accordingly. This classification task aims to constrain the feature distances of all the positive pairs less than those of all the negative pairs, regardless of whether these pairs correspond to the same query image. With this stronger constraint than the triple ranking task, the classification task shall pull the positive image pairs with less appearance similarity closer. Meanwhile, the negative image pairs with larger appearance similarity are pushed farther. This brings in compact intra-place feature distribution and separable inter-place feature distribution. It is worth mentioning that only using the classification task will impair the order of similarity in appearance, and thus, the proposed binary classification task is combined with the existing triplet ranking task to promote each other. As a result, the generalization of our model is enhanced. Secondly, an attention module is designed to improve the pooling layer of NetVLAD (VLAD network) [19]. By capturing flexible context information and regularizing the distribution of learned feature map, this module assigns different importance to each spatial position during feature aggregation for better image global feature extraction.

The rest of the chapter is structured as follows. Section 5.2 presents the related work. The proposed multi-task learning method is elaborated in Section 5.3. Experiments are shown in Sections 5.4, and 5.5 concludes this chapter.

5.2 REVIEW OF VISUAL PLACE RECOGNITION METHODS

There are many works dedicated to visual place recognition for reliable localization. The traditional methods resort to local or global handcrafted features to describe places. The local descriptors can be divided into floating-point and binary types. The former include

SIFT [2] and speeded-up robust features (SURF) [20]. Binary robust independent elementary features (BRIEF) [21] and binary robust invariant scalable keypoints [22] are two typical descriptors of the latter. These extracted local descriptors are then aggregated into a holistic image descriptor using aggregation techniques such as BoW [3], VLAD [4], or fisher vectors [23], which provide the foundation of place recognition. Note that each local descriptor follows the procedure of keypoint detection and description. In contrast, the global descriptors such as histogram of oriented gradient (HOG) [24] dismiss the detection step and directly process the whole image. An impressive case is fast appearance-based mapping (FAB-MAP) [25], which extracts the SURF features from an image and converts them into a BoW representation for effective place recognition. Subsequent works make improvements on BoW-based place recognizers, either with features from accelerated segment test (FAST) keypoints and BRIEF descriptors for efficiency [26] or with binary oriented FAST and rotated BRIEF (ORB) features for both efficiency and robustness [27].

With the successful applications of CNNs on computer vision tasks [28,29], more and more researches adopt CNNs to extract effective image representation for place recognition. Similar to traditional local descriptor-based solutions, the CNN feature map can be regarded as dense local features and the feature of each spatial position captures the local information in the neighborhood. Benefiting from the learnable features under constraint of loss functions, the CNN-based place recognition is thought to be more robust and discriminative [6,30]. A typical pioneering example is NetVLAD [19], where a generalized VLAD layer is proposed to aggregate the feature map from the CNN backbone into a global image feature. This VLAD layer can be easily plugged into a CNN architecture for end-to-end training.

NetVLAD treats each feature in the feature map equally in the aggregation process; however, the importance of each position is actually different. Some landmarks such as buildings are beneficial to recognize a place while some confusing visual elements or even dynamic objects can bring in perturbation [16]. To solve this problem, researchers focus on various attention mechanisms. Kim et al. [14] introduced a contextual reweighting network (CRN) to estimate the importance of each spatial position in the feature map, which guides the model to concern the salient regions for effective place recognition. A multi-scale context-flexible attention model is built for saliency estimation [15], where multiple soft attention masks are predicted from different convolution layers followed by multi-scale attention fusion. In reference [5], a second-order attention module is employed and combined with a NetVLAD layer for improved global features. The above methods adopted various strategies to capture contextual information; however, the influence of image appearance variations on features is seldom considered. In contrast to the standard convolutions in CRN [14], dilated convolutions are employed in our method to capture more context information. Furthermore, domain normalization (DN) operations are appended to the four dilated convolutions and the robustness of features to environment changes is improved.

To organize the image representations well, the loss function is crucial. It aims to reduce the feature distances between a query image and its positive images, meanwhile, enlarge those between a query image and its negative samples. A quintuplet loss is proposed by Zhai et al. [7], which embeds all the potential positive samples into the primitive triplet loss to emphasize the constraints between anchor samples and positive samples. Liu et al. [9] improved the triplet loss of NetVLAD by introducing a new stochastic attraction and repulsion embedding (SARE) loss function for maximizing the ratio between the query-positive pair against multiple query-negative pairs. On the basis of [9], self-supervised fine-grained region similarity (SFRS) [31] is presented to achieve better feature learning, where the image-to-region similarities are considered to fully explore the potential of difficult positive images. The spatial pyramid-enhanced VLAD (SPE-VLAD) with a weighted triplet loss (WT-loss) is proposed [8]. It imposes temporal constraints among the features between adjacent epochs, inhibiting the increase of feature distances for the positive sample pairs in the adjacent epochs. Different from these methods, we increase a classification task to achieve more compact intra-place features and more separable inter-place features.

5.3 VISUAL PLACE RECOGNITION VIA A MULTI-TASK LEARNING METHOD WITH ATTENTIVE FEATURE AGGREGATION

The proposed multi-task learning network with attentive feature aggregation, termed as MTA, aims to learn a good global image representation for visual place recognition, which is described in Figure 5.1. On this basis, following the procedure of [19], we perform dimension reduction on the global image feature, and then the place of the query image can be recognized from the reference images through a feature-based nearest neighborhood retrieval. Specifically, the multi-task network mainly consists of three modules: local feature extraction, attentive feature aggregation, and a binary classification network, where the first two modules constitute the global feature extraction network. Note that the proposed network is trained by jointly optimizing a triplet ranking task and a binary classification task, and the global image features can be learned end to end.

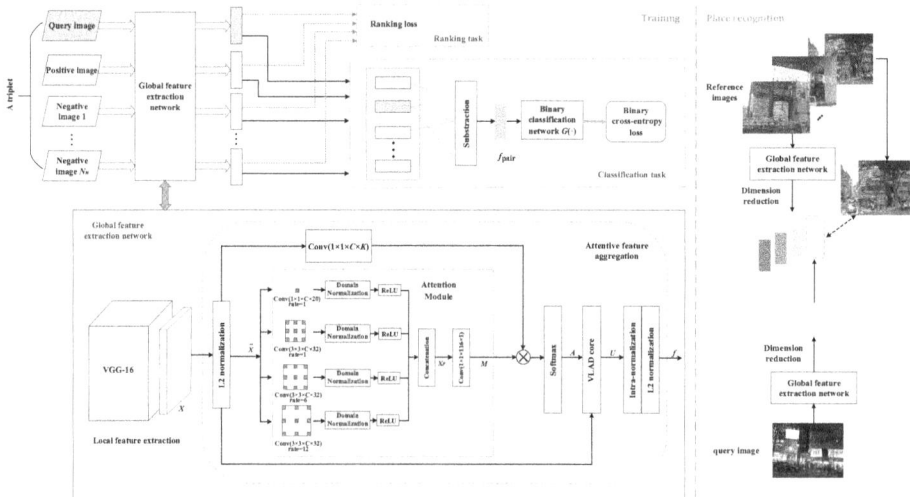

FIGURE 5.1   The pipeline of the proposed multi-task learning method with attentive feature aggregation for visual place recognition.

5.3.1 Local Feature Extraction with CNN

CNN is generally taken as a visual encoder to capture the local contextual features from an input image. We use VGG-16 (visual geometry group) truncated after conv5_3 as the backbone [9,19,31] to encode the features of input images. The weights of the encoder are initialized with the VGG-16 pre-trained on ImageNet dataset [32], and all the initial parameters before conv5 are frozen to preserve the robustness to environment changes [16]. The feature map $X \in R^{H \times W \times C}$ outputted from the encoder Enc(·) can be denoted as

$$X = \text{Enc}_{\theta_1}(I) \tag{5.1}$$

where I refers to an input image, and H, W, and C are the height, width, and channel number of X, respectively. θ_1 denotes the encoder parameter set. In essence, the feature map records a set of C-channel local features at $H \times W$ spatial positions.

5.3.2 Attentive Feature Aggregation

The attentive feature aggregation module serves to congregate the extracted local features in a single image into a compact global feature vector. It is well known that different regions in an image do not contribute equally to describe a place. Therefore, an attention module is designed to highlight the important regions by learning a $H \times W \times 1$ dimensional spatial attention map. This attention module is embedded into the VLAD layer [19] and a weighted feature aggregation scheme is accomplished. In addition, an L_2 normalization operation in channel is performed at the beginning of feature aggregation to expedite the convergence of network.

5.3.2.1 Attention Module

The attention module is operated on the L_2 normalized feature map, which is denoted as $\bar{X} \in R^{H \times W \times C}$. The architecture of this module is exhibited in Figure 5.1, and it utilizes four context filters in parallel to capture multi-scale context information. These filters are implemented by four convolution layers with different configurations. Dilated convolution [33] is involved to ensure a larger receptive field with less model parameters. Concretely, a 1×1 convolution and three 3×3 convolutions with dilated rates of 1, 1, 6, and 12 are applied to the normalized feature map $\bar{X}$. The kernel size and dilated rate of a filter control the size of local neighborhood that can be directly perceived at each spatial location. Since different objects in an image have different sizes, the application of multi-scale context filters with dilated convolutions tends to obtain semantically stronger features with better discrimination. Besides, the proposed network is required to be robust to environmental changes such as illumination and season. The DN operation [34] used in the field of stereo matching provides a preferable solution. It is implemented via an instance normalization followed by an L_2 normalization along the channel dimension and a linear transformation with learnable scale and shift parameters. For each element x_i in the input feature map, instance normalization is first employed to normalize features along the spatial dimension, and $\hat{x}_i$ is generated, then L_2 normalization in channel and linear transformation are used as follows [34]:

$$y_i = \frac{\hat{x}_i}{\sqrt{\sum_{j \in S_i} \left|\hat{x}_j\right|^2 + \epsilon}} \gamma_i + \beta_i \tag{5.2}$$

where S_i represents the indexes of elements across C channels with the same spatial position as $\hat{x}_i$. γ_i and β_i refer to scale and shift parameters, respectively. All y_i constitute the output of a DN module.

The instance normalization helps to reduce the influence of image-level style variations and L_2 normalization emphasizes the local invariance. In this chapter, DN operation is appended to each context filter to regulate the distribution of features. Afterward, the non-linear activation function ReLU (rectified linear units) is exerted to generate corresponding feature maps. Finally, the four feature maps are concatenated in channel to form the fused feature map X_F. A 1×1 convolutional layer with weight K_l and bias b_l is employed on X_F and we obtain final spatial attention map $M \in R^{H \times W \times 1}$. Formally,

$$X_F = \bigcup_{p=1}^{4} \left[g_p \left(K_p \otimes \overline{X} \oplus b_p \right) \right] \tag{5.3}$$

$$M = K_l \otimes X_F \oplus b_l \tag{5.4}$$

where $\otimes$ and $\oplus$ constitute convolution operation and $\cup$ means the concatenation operator. K_p and b_p refer to the weight and bias of pth context filter, where $p = 1, \ldots, 4$. g_p stands for the operation of DN with ReLU.

5.3.2.2 Feature Aggregation

As shown in Figure 5.1, the generated spatial attention map M is embedded into the original VLAD layer [19] to achieve an attentive feature aggregation. Herein, we continue to use the visual dictionary built in NetVLAD, which is composed of K clusters and the feature vector of each cluster center is denoted as $c_k \in R^C$, $k = 1, 2, \ldots, K$. NetVLAD imposes a convolution on the feature map X followed by softmax activation to obtain a soft assignment matrix. In contrast, we take the normalized feature map $\overline{X}$ as the input and an additional attention branch is also involved. Thus, not only the soft assignment coefficient is considered, but also the importance of feature for each spatial position before aggregating them is emphasized. In practice, an element-wise multiplication is executed across each channel of the output of convolution $\text{Conv}(1 \times 1 \times C \times K)$ with attention map M, and then the soft assignment matrix can be obtained as follows:

$$A = \text{Softmax}\left[\left(K_a \otimes \overline{X} \oplus b_a \right) * M \right] \tag{5.5}$$

where K_a and b_a are the weight and bias of $\text{Conv}(1 \times 1 \times C \times K)$, respectively. $A \in R^{H \times W \times K}$ and its each element A_{hwk} manifests to what extent the local feature $\bar{X}_{hw} \in R^C$ at spatial location (H, W) of feature map $\bar{X}$ belongs to the kth cluster. Then the local feature set $\{\bar{X}_{hw}\}$ is assembled into a feature matrix $U \in R^{C \times K}$ by the VLAD core as follows:

$$U = [u_1, \ldots, u_k, \ldots, u_K] \tag{5.6}$$

where $u_k = \sum_{h=1}^{H} \sum_{w=1}^{W} A_{hwk}(\bar{X}_{hw} - c_k)$. After the matrix U passes through intra-normalization, it is flattened into a feature vector. Applying L_2 normalization to this vector and we acquire the final global feature representation f.

5.3.3 Triplet Ranking Task

For visual place recognition, a good feature representation is desired so that the location corresponding to the query image can be speculated by finding the matched reference images in the feature space. Such a case is usually processed through a triplet ranking task.

To facilitate the learning of ranking task, one needs to find out the suitable positive and negative samples for each query image, which constitute a triplet. Such a triplet is determined according to the geographical distances and feature distances between a query image and reference images [19,31]. Given a query image, the reference images within its certain geographical distance are regarded as the candidate positive samples, among which the sample with the smallest feature distance to the query image is regarded as the final positive sample. After the candidate negative samples are confirmed according to geographical distance, a small amount of hard negative samples are selected [19]. Then, we feed triplets into the network for training, each of which contains 1 query image, a positive image, and N_n negative images. The objective of triplet ranking task is to enforce the feature similarity between f_{q_i} and f_{p_i} to be larger than that between f_{q_i} and $f_{n_i^j}$, where f_{q_i} and f_{p_i} are the global features of ith query image and its positive image, respectively. $f_{n_i^j}$ refers to the global feature of jth negative image related to i^{th} query image. The ranking loss in reference [31] is adopted as follows:

$$L_r = -\sum_{i=1}^{N_q} \sum_{j=1}^{N_n} \log \frac{\exp\left(\frac{S(f_{q_i}, f_{p_i})}{t_e}\right)}{\exp\left(\frac{S(f_{q_i}, f_{p_i})}{t_e}\right) + \exp\left(\frac{S(f_{q_i}, f_{n_i^j})}{t_e}\right)} \tag{5.7}$$

where N_q is the number of query images and t_e is a temperature parameter. $S(f_1, f_2) = f_1^T f_2$ is a cosine similarity function.

To make a further analysis, we consider that q_i, p_i, n_i^j only exist in the ith triplet, and the gradients of this objective loss with respect to f_{p_i}, $f_{n_i^j}$, and f_{q_i} are computed as follows:

$$\frac{\partial L_r}{\partial f_{p_i}} = -\sum_{j=1}^{N_n} \frac{\exp\left(\frac{f_{q_i}^T f_{n_i^j}}{t_e}\right)}{\exp\left(\frac{f_{q_i}^T f_{p_i}}{t_e}\right) + \exp\left(\frac{f_{q_i}^T f_{n_i^j}}{t_e}\right)} \frac{f_{q_i}}{t_e} \tag{5.8}$$

$$\frac{\partial L_r}{\partial f_{n_i^j}} = \frac{\exp\left(\frac{f_{q_i}^T f_{n_i^j}}{t_e}\right)}{\exp\left(\frac{f_{q_i}^T f_{p_i}}{t_e}\right) + \exp\left(\frac{f_{q_i}^T f_{n_i^j}}{t_e}\right)} \frac{f_{q_i}}{t_e} \tag{5.9}$$

$$\frac{\partial L_r}{\partial f_{q_i}} = \sum_{j=1}^{N_n} \frac{\exp\left(\frac{f_{q_i}^T f_{n_i^j}}{t_e}\right)}{\exp\left(\frac{f_{q_i}^T f_{p_i}}{t_e}\right) + \exp\left(\frac{f_{q_i}^T f_{n_i^j}}{t_e}\right)} \frac{f_{n_i^j} - f_{p_i}}{t_e} \tag{5.10}$$

One can see from (5.8) to (5.10) that the gradients $\frac{\partial L_r}{\partial f_{q_i}}$, $\frac{\partial L_r}{\partial f_{p_i}}$, and $\frac{\partial L_r}{\partial f_{n_i^j}}$ of the ranking loss depends only on the ith triplet of the training set. The network is trained with stochastic gradient descent optimizer that computes the gradient from individual triplets, which means that the global structure of the dataset is not concerned and the network is prone to overfitting [35,36]. For the network training with only triplet ranking task, increasing the mini-batch size is helpful to relieve the overfitting problem of triplet ranking loss. This is due to the fact that when the number of triplets in a batch increases, the network can be better trained, considering more image features inside the same batch. Even so, such process individually constrains the samples in each batch and the correlation among different batches is not considered, which leads to that the improvement of overfitting is limited. Another way to improve overfitting is to add normalization layers, which conduct the normalization operation on each batch of training data. It is beneficial to adjust the data distribution in a batch; however, it still neglects the correlation among different batches.

5.3.4 Binary Classification Task

In this section, a binary classification task is added to constrain the feature distances of all the positive pairs less than those of all the negative pairs, regardless of whether these pairs are from the same triplet/batch. We focus on the paired images selected from the ranking task instead of separately processing them. The query-positive pairs belong to positive class and query-negative pairs constitute negative class. The binary classification task consists of a binary classification network and a corresponding binary cross-entropy loss.

5.3.4.1 Binary Classification Network

As can be seen from Figure 5.1, the difference f_{pair} of features for each paired images is regarded as the input of the classification network:

$$f_{\text{pair}} = f_o - f_q \qquad (5.11)$$

where f_q and f_o are the query feature and positive/negative feature, respectively.

The structure of binary classification network $G(\cdot)$ is shown in Figure 5.2, which includes two fully connected layers with nonlinear functions, where dropout serves to avoid over-fitting and leaky ReLU retains the input information with negative values. The network outputs a probability p that the corresponding paired images belong to the positive class:

$$p = G_{\theta_2}\left(f_{\text{pair}}\right) \qquad (5.12)$$

where θ_2 is the parameter set of the classification network.

5.3.4.2 Binary Cross-Entropy Loss

Given the above binary classification task, we employ a binary cross-entropy loss to train the network, which is formulated as

$$L_c = -\sum_{l=1}^{L}\left[y_l \log p_l + \left(1 - y_l\right)\log\left(1 - p_l\right)\right] \qquad (5.13)$$

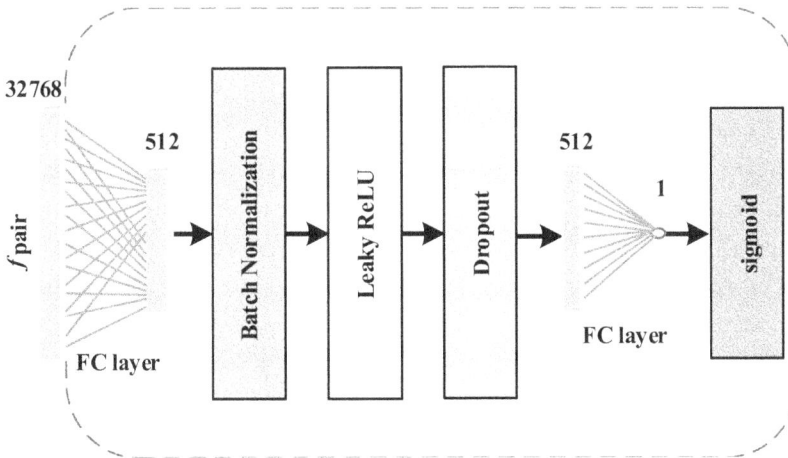

FIGURE 5.2   The structure of binary classification network.

where p_l and y_l describe the output probability and label of lth pair, respectively. $l = 1, 2, \ldots, L$ and $L = N_q(1 + N_n)$ is the number of pairs. When a pair is a query-positive one, $y_l = 1$, or else $y_l = 0$. More specifically,

$$
\begin{aligned}
L_c = &-\sum_{i=1}^{N_q} \log\left(G_{\theta_2}\left(f_{p_i} - f_{q_i}\right)\right) \\
&-\sum_{i=1}^{N_q}\sum_{j=1}^{N_n} \log\left(1 - G_{\theta_2}\left(f_{n_i^j} - f_{q_i}\right)\right)
\end{aligned}
\tag{5.14}
$$

The gradients of binary cross-entropy loss in (14) with respect to f_{p_i}, $f_{n_i^j}$, and f_{q_i} can be derived as follows:

$$
\frac{\partial L_c}{\partial f_{p_i}} = -\frac{G'_{\theta_2}\left(f_{p_i} - f_{q_i}\right)}{G_{\theta_2}\left(f_{p_i} - f_{q_i}\right)}
\tag{5.15}
$$

$$
\frac{\partial L_c}{\partial f_{n_i^j}} = \frac{G'_{\theta_2}\left(f_{n_i^j} - f_{q_i}\right)}{1 - G_{\theta_2}\left(f_{n_i^j} - f_{q_i}\right)}
\tag{5.16}
$$

$$
\frac{\partial L_c}{\partial f_{q_i}} = \frac{G'_{\theta_2}\left(f_{p_i} - f_{q_i}\right)}{G_{\theta_2}\left(f_{p_i} - f_{q_i}\right)} - \frac{G'_{\theta_2}\left(f_{n_i^j} - f_{q_i}\right)}{1 - G_{\theta_2}\left(f_{n_i^j} - f_{q_i}\right)}
\tag{5.17}
$$

It is seen from (5.15) to (5.17) that the gradients $\dfrac{\partial L_c}{\partial f_{q_i}}$, $\dfrac{\partial L_c}{\partial f_{p_i}}$, and $\dfrac{\partial L_c}{\partial f_{n_i^j}}$ for the binary cross-entropy loss depend not only on the ith triplet but also on the parameter set θ_2 of the designed binary classification network. The parameter set θ_2 is continuously updated according to the input pair features. Specifically, two fully connected layers are involved in the binary classification network, and their parameters are denoted as $W_1 \in R^{32,768 \times 512}$ and $W_2 \in R^{512 \times 1}$, respectively. We denote with $W = W_1 W_2 \in R^{32,768 \times 1}$. Therefore, $W_1 \subset \theta_2$, $W_2 \subset \theta_2$, and $W \subset \theta_2$. Since the label of each positive pair is 1, the result of dot product between W and the input feature $f_{\text{pair}}^p \in R^{32,768 \times 1}$ of each positive pair is forced to be as large as possible, which implies that W and f_{pair}^p are similar enough. As a result, W can be treated as the center of the positive class, which actually reflects the whole feature distribution of positive pairs in all triplets of the whole training dataset. When the feature of each sample in a triplet is updated according to (5.15)–(5.17), W is involved, and the global statistics are considered. This helps to suppress the interference of outliers and thus the overfitting is effectively prohibited. It is worth mentioning that the binary classification network is only used for the training process to promote feature learning and it shall be dropped at testing phase. Thus, it does not increase computation burden once the model is deployed.

With the above ranking loss and classification loss, the final joint training objective L is given by

$$L = L_r + \alpha L_c \qquad (5.18)$$

where α is a weight to balance the losses of two tasks.

The first term L_r is a pair-based loss, which captures the fine-grained paired semantic relations and thus facilitates the instance discrimination. However, it is inclined to slow convergence and local optimization. The second term L_c regulates the distances among features and thus improve the generalization of model. These two related tasks interact with each other and a robust global feature representation is learned.

5.3.5 Visual Place Recognition via Feature-Based Nearest Neighborhood Retrieval

The dimension of the learned global feature is large, which shall increase the computation cost and thus slow down the place recognition. A principal component analysis (PCA) module is adopted to reduce the dimension of global features for efficient place recognition. After the global feature extraction network is trained, the parameters of PCA are then updated according to the global features of all training images. The trained PCA module is appended to the global feature extraction network, and the global feature with a smaller dimension is obtained. The original global feature is compressed with the dimension from $32,768 \times 1$ to $4,096 \times 1$. For a test scene, we pre-computed the compressed global features of reference images. Each query image is inputted into the trained model to generate a compressed global feature. Next, the Euclidean distances between the query feature and all reference features are computed, and the reference image with minimum feature distance is selected.

5.4 EXPERIMENTS

In this section, the proposed MTA method is verified by experiments on public datasets and actual environments.

5.4.1 Experimental Setup

5.4.1.1 Datasets

Following state-of-the-art visual place recognition methods [7,9,19,31], the model is trained on the Pittsburgh Pitts30k dataset [37]. The complete Pittsburgh dataset called Pitts250k consists of 250 k reference images and 24 k query images, where each image is annotated with its global positioning system (GPS) coordinate. Notice that the query images and reference images are captured at different times (or even years apart). Also, there are multiple perspective images with different viewpoints at each GPS coordinate. This dataset actually covers various viewpoints, illumination, and season conditions. Pitts30k is a subset of Pitts250k, and it includes 30 k reference images and 22 k query images. For Pitts30k and Pitts250k datasets, the query and reference images are both divided into roughly equal training, validation, and test parts.

To validate the performance of the proposed global feature extraction network that is trained on Pitts30k dataset, several place recognition datasets are used for testing:

TokyoTM-val [19], Pitts250k-test [37], and Tokyo 24/7 [38]. TokyoTM-val dataset is constructed using the images from Google street view of Tokyo. Tokyo 24/7 dataset contains 76 k reference images and 315 query images. The reference images are attained at daytime while the query images are obtained at various conditions such as daytime, sunset, and night, which makes Tokyo 24/7 more challenging. Besides, an image retrieval dataset (Holidays dataset) is considered with a set of images of various high resolutions, which is divided into 500 query images and 991 reference images.

5.4.1.2 Evaluation Metric
To evaluate the place recognition performance, the common metric of top-k recall is adopted. Each query image is considered to be correctly localized, when at least one of the first k similar reference images is located within certain geographical distance from the query image. The geographical distance threshold is set to 25 m [19,31] for Pitts250k-test dataset and Tokyo 24/7 dataset, 10 m [8] for the TokyoTM-val dataset. The percentage of query images that are correctly localized is calculated as the top-k recall. As for the performance evaluation for image retrieval, the mean average precision (mAP) is employed.

5.4.1.3 Implementation Details
Our method is implemented on the platform of NVIDIA GeForce GTX 1080 Ti GPU and Intel Xeon E5-2650 v4 CPU. At the training phase, random ColorJitter is applied on the training images for data augmentation. The batch size of triplets for training is set to 4. We train our model end to end for 5 epochs, and the model with the largest top-5 recall on Pitts30k validation set is deemed as the final model for place recognition. The stochastic gradient descent optimizer is used with momentum and weight decay of 0.9 and 0.001, respectively. The initial learning rate is 0.001, which is reduced by half every two epochs.

5.4.2 Ablation Studies
To testify the performance of our MTA method, its six variants, MTA_I, MTA_II, MTA_III, MTA_IV, MTA_V, and MTA_VI, are considered according to whether ranking loss, classification loss, context filters with standard convolutions, the proposed context filters with dilated convolutions, and DN are considered, where the last two items are main components of our attention module. The context filters with standard convolutions are achieved by replacing the 3×3 dilated convolutions using standard convolutions with kernels 3×3, 5×5, and 7×7. Table 5.1 presents the comparison results on Pitts250k-test and Tokyo 24/7 datasets in terms of top-1, top-5, and top-10 recalls. The best results are labeled in bold. The results of MTA_I and MTA_II show the performance of individual ranking task and classification task, respectively, which indicates that ranking loss is superior to classification loss. Combining these two losses, MTA_III yields a better place recognition performance. With the help of context filters with dilated convolutions, the performance is further improved (see MTA_IV). From the results of MTA, the combination of our context filter and DN with two losses performs well. MTA performs better than MTA_IV especially on the challenging Tokyo 24/7 dataset with severe illumination variations, which demonstrates the effectiveness of DN. Also, the comparison of MTA_V and MTA verifies the generalization of the

TABLE 5.1 Comparison of Different Variants of Our MTA Method on Place Recognition Datasets in Terms of Recall (%)

Method	Ranking Loss	Classification Loss	Context Filters with Standard Convolutions	Context Filters with Dilated Convolutions	Domain Normalization	Pitts250k-test			Tokyo 24/7		
						Recall@1	Recall@5	Recall@10	Recall@1	Recall@5	Recall@10
MTA_I	✓	—	—	—	—	89.5	95.4	96.8	81.6	88.6	90.8
MTA_II	—	✓	—	—	—	87.1	94.0	95.7	77.1	84.4	87.9
MTA_III	✓	✓	—	—	—	89.5	95.7	97.1	80.6	89.8	93.0
MTA_IV	✓	✓	—	✓	—	90.5	95.9	97.1	85.1	91.1	93.7
MTA_V	✓	—	—	✓	✓	**90.8**	**96.1**	97.2	83.5	91.4	93.0
MTA_VI	✓	✓	✓	—	✓	90.0	95.8	97.0	**85.7**	90.2	92.1
MTA	✓	✓	—	✓	✓	90.4	**96.1**	**97.3**	**85.7**	**92.7**	**94.6**

TABLE 5.2   Comparison of Different Methods on TokyoTM-val Dataset in Terms of Recall (%)

Method	Recall@2	Recall@3	Recall@4	Recall@5	Recall@10	Recall@15	Recall@20	Recall@25
SPE-VLAD+T-loss [11]	19.93	37.14	50.46	59.48	76.43	81.47	83.77	85.47
MH-SA+WT-loss	17.02	34.23	48.18	56.62	75.06	80.11	82.90	84.62
NetVLAD+WT-loss	24.35	42.97	54.48	61.58	75.13	79.72	82.00	83.75
SPE-VLAD [11]	23.20	41.98	55.15	63.90	77.28	81.54	83.97	85.61
DenseVLAD [41]	84.00	86.81	88.54	89.44	91.80	92.69	93.29	93.69
NetVLAD [22]	91.30	93.22	94.18	94.81	96.03	96.62	96.90	97.11
AP-GEM [42]	83.02	86.03	87.58	88.84	91.51	92.77	93.54	94.11
SARE [12]	92.02	94.00	94.84	95.37	96.48	96.95	97.27	97.43
SFRS [34]	**92.54**	94.35	95.16	95.60	96.74	97.22	97.43	97.64
MTA (Ours)	92.50	**94.40**	**95.25**	**95.81**	**96.93**	**97.38**	**97.65**	**97.81**

classification task. The results of MTA_VI and MTA imply that dilated convolution is better to capture flexible context information than the common standard convolution.

Next, the execution speed of the model is given. Take the Pitts250k-test dataset as an example. The average execution time of global feature extraction and PCA dimension reduction is about 34 ms per frame. Correspondingly, the execution speed of the model is 29.4 FPS. If the time of nearest neighborhood retrieval is counted, the increasing execution time is positively related to the number of reference images. With 10,000 reference images, the average retrieval time of a query image is about 46 ms. Thus, the total time of place recognition for a query image is 80 ms and the speed is 12.5 FPS.

5.4.3 Comparison with Existing Methods

In this section, the proposed MTA method is compared with existing methods on place recognition datasets. Table 5.2 provides the results of the TokyoTM-val dataset in terms of different recalls. The methods include SPE-VLAD [8], DenseVLAD [38], NetVLAD [19], average precision-generalized mean pooling (AP-GEM) [39], SARE [9], and SFRS [31]. DenseVLAD is a traditional feature representation method that extracts dense SIFT features and congregates them with intra-normalized VLAD. The others belong to CNN-based solutions, where AP-GEM aggregates local CNN features by a generalized mean pooling layer and the rest are NetVLAD and its improved versions. The results of NetVLAD and SARE methods in Table 5.2 are acquired using the trained model provided in [40,41], respectively. One can see that SFRS and our method achieve the best performance. Compared to SFRS, MTA is slightly better. Take two query images from the TokyoTM-val dataset as examples, and the top-3 retrieved results of the methods including NetVLAD, SARE, SFRS, and MTA are illustrated in Figure 5.3, where two query images are located in the first column of Figure 5.3a and b, respectively. For the query image shown in the first column of Figure 5.3a, all retrieved results of all methods are correct. Obviously, large common salient regions between the query image and reference images are beneficial to place recognition. On the contrary, ambiguous regions shall cause interference with the task, which can be seen from the results of Figure 5.3b. NetVLAD and SARE fail to recognize the query image, while SFRS only acquires the correct result on the second retrieved image.

(a)

(b)

FIGURE 5.3 Visualization of the top-3 retrieved images of different methods for two selected query images from TokyoTM-val dataset. (a, b) The query image and the top-3 retrieved results, which are presented in the first column and the last three columns, respectively.

TABLE 5.3 Comparison of Different Methods on Place Recognition Datasets in Terms of Recall (%)

	Pitts250k-test			Tokyo 24/7		
Method	Recall@1	Recall@5	Recall@10	Recall@1	Recall@5	Recall@10
NetVLAD [19]	85.95	93.20	95.13	73.33	82.86	86.03
CRN [14]	85.50	93.50	95.50	75.20	83.80	87.30
SARE [9]	88.97	95.50	96.79	79.68	86.67	90.48
SFRS [31]	**90.68**	**96.39**	**97.57**	85.40	91.11	93.33
APANet [42]	83.65	92.56	94.70	66.98	80.95	83.81
QUITLoss [7]	84.03	92.16	94.01	69.52	81.59	84.76
AP-GEM [39]	79.94	90.83	93.55	40.3	55.6	65.4
Dense VLAD [38]	76.82	84.95	86.17	59.4	67.3	72.1
MTA (Ours)	90.43	96.11	97.27	**85.71**	**92.70**	**94.60**

For our method, the top-3 images are valid, which demonstrates the advantage of our multi-task learning with attentive feature aggregation. In addition, the top-2 images of NetVLAD as well as the top-1 images of SARE and SFRS are actually similar to the query image; however, they are regarded as the wrong results as the constraint of geographical distance threshold is not satisfied. Compared to the top-3 retrieval results of MTA, the viewpoints of these four retrieved images are more similar to the query viewpoint, which indicates that our proposed method is insensitive to viewpoints.

Table 5.3 lists the comparison of different methods on Pitts250k-test and Tokyo 24/7 datasets in terms of recalls. DenseVLAD [38], NetVLAD [19], attention-based pyramid aggregation network (APANet) [42], AP-GEM [39], quintuplet loss (QUITLoss) [7], CRN [14], SARE [9], and SFRS [31] are involved to compare with our MTA method. As is shown in Table 5.3, the proposed MTA method and SFRS still take the top-2 spots. Our method is 0.25%, 0.28%, and 0.30% lower than SFRS in terms of top-1, top-5, and top-10 recalls on the Pitts250k-test dataset, whereas on Tokyo 24/7 dataset, it is 0.31%, 1.59%, and 1.27% higher than SFRS. Besides, SFRS needs to train the model with a self-supervised enhancement for four generations and each generation includes five epochs. In contrast, MTA is trained only five epochs with a small amount of training time. These results manifest the effectiveness of our method.

Take two query images from Pitts250k-test dataset as examples, and the top-3 retrieved results of NetVLAD, SARE, SFRS, and our MTA are displayed in Figure 5.4. For the query image in the first column of Figure 5.4a, SFRS succeeds to retrieve the top-3 reference images and our method fails in the third retrieval result. One can see that the right half region of the third retrieval image for SFRS has a larger similarity with the query image. Such a case can be correctly retrieved because SFRS considers the image-to-region similarity during training, which makes it focus on the local features. By contrast, our MTA mainly concerns the image-level feature, and sometimes partial image details are neglected. For the query image in Figure 5.4b, it contains repetitive texture, which is confusing for place recognition. Both SFRS and MTA retrieve a wrong image while the retrieval results of NetVLAD and SARE are all wrong.

Figure 5.5 illustrates the top-1 retrieved images of different methods on Tokyo 24/7 dataset for three query images. These query images are taken at day time, sunset, and night

(a)

(b)

FIGURE 5.4 Visualization of the top-3 retrieved images of different methods for two selected query images from Pitts250k-test dataset. (a, b) The top-3 retrieved results of two query images, respectively.

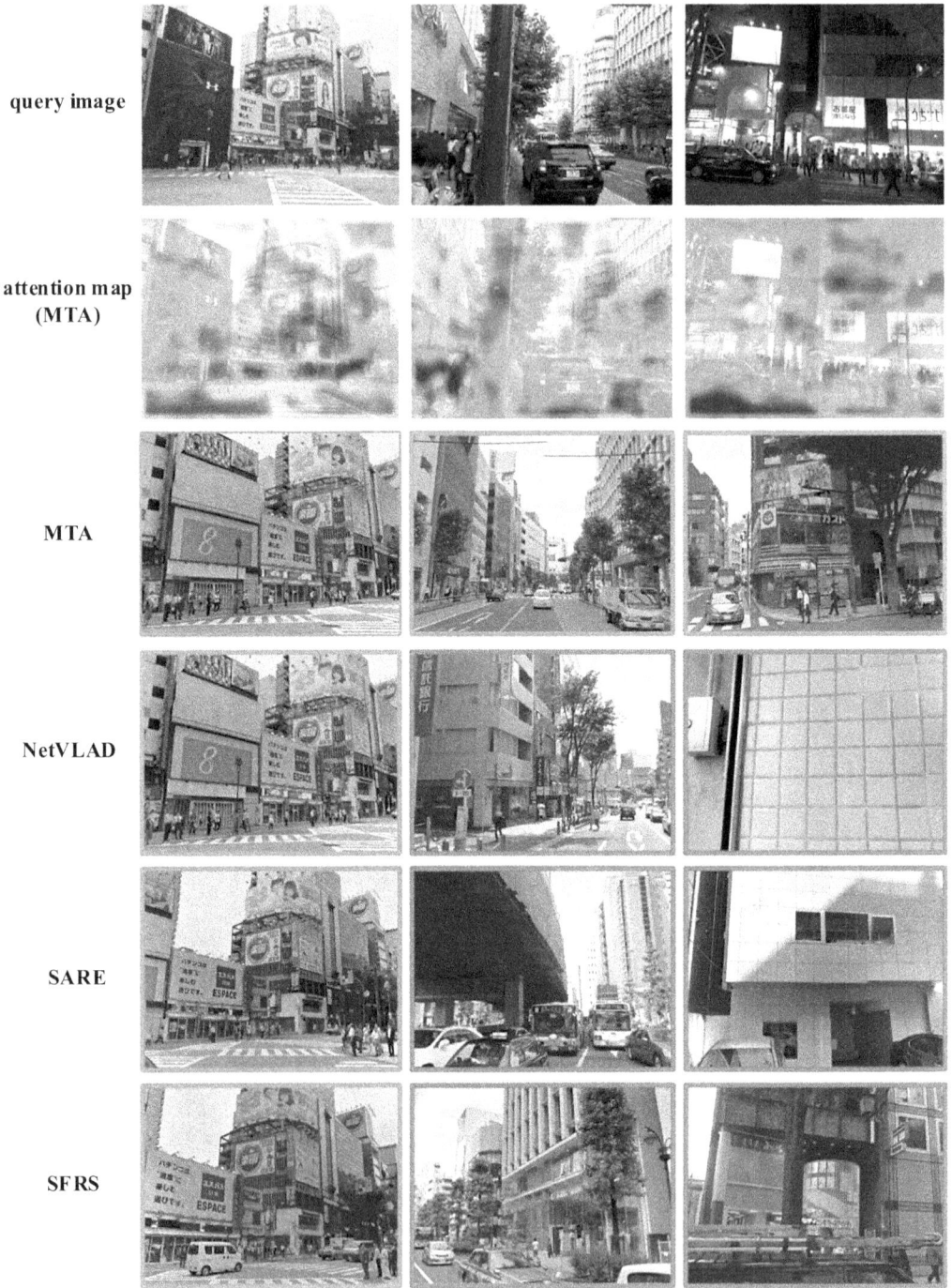

FIGURE 5.5 Visualization of the top-1 retrieved images of different methods for three selected query images from Tokyo 24/7 dataset. The attention map of our method for each query image is also provided in Line 2. The query image of each column corresponds to day time, sunset, and night, respectively.

with confusing objects such as trees and dynamic pedestrian. The attention map generated from our MTA method is also visualized. One can see from the third line of Figure 5.5 that our method achieves correct retrieval, which implies the advantage of our method. The attention maps also reflect that our proposed attention module can pay more attention to discriminative buildings, street lamps, and poles while ignoring those confusing objects. To further explore the robustness of our method, we split all the query images of the Tokyo 24/7 dataset into two subsets: day time queries and sunset/night queries. The recall curves of our method, SARE, and SFRS on these two subsets are described in Figure 5.6a, where each curve records the top-N recall for different values of N. The results indicate that the three methods are basically similar on day time queries, and our method is better on sunset/night queries. This phenomenon manifests the robustness of our method. Considering the performance of all query images in Figure 5.6b, our method is considered as an effective one.

5.4.4 Generalization on Image Retrieval Dataset

To further assess the generalization of our proposed multi-task learning method, we directly deploy our global feature extraction network trained on Pitts-30k dataset to the Holidays dataset [43] for global feature extraction. Table 5.4 lists the retrieval results of our method and three state-of-art methods. One can conclude that MTA reaches the best performance.

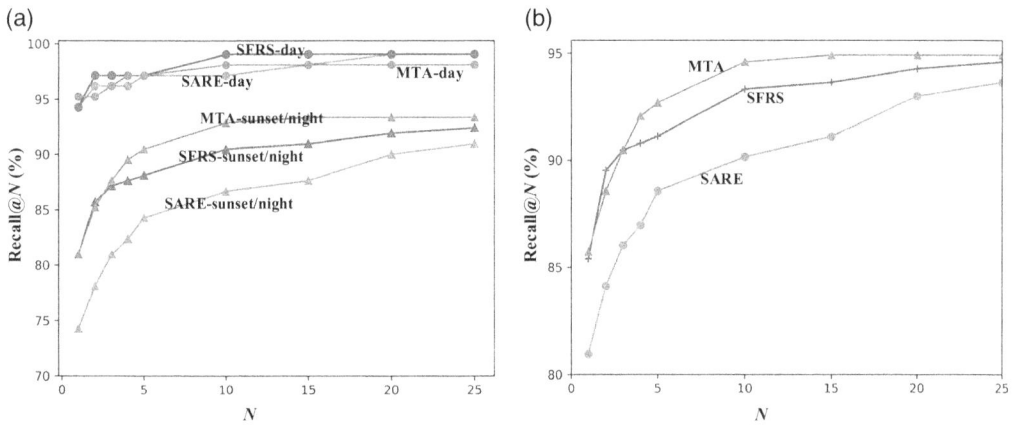

FIGURE 5.6   The recall curves of different methods in terms of top-N recall on Tokyo 24/7 dataset. (a) Results of day time queries and sunset/night queries. (b) Results of all queries.

TABLE 5.4   Comparison of Different Methods on Holidays Dataset in Terms of mAP (%)

Methods	NetVLAD	SARE	SFRS	MTA
mAP	83.0	80.7	80.5	**83.8**

Taking two images selected from Holidays dataset as examples, top-3 retrieved images of SFRS and our MTA for them are displayed in Figure 5.7. For the query image in Figure 5.7a, our MTA retrieves two correct reference images, whereas SFRS only recognizes one correct result among the top-3 retrieved images. Both methods have the same top-2 retrieval results and the second image is mistakenly recognized, which might be due to the similar local structure to the query image. The third retrieval result of SFRS has quite large difference from the query image while the one for our MTA is correct. Although the sky and cloud bring in the interference on the feature representation, MTA is still able to locate the salient region for reliable retrieval. From the results in Figure 5.7b, our method retrieves three correct reference images while SFRS fails in the third retrieval result, which demonstrates MTA could extract discriminative global features.

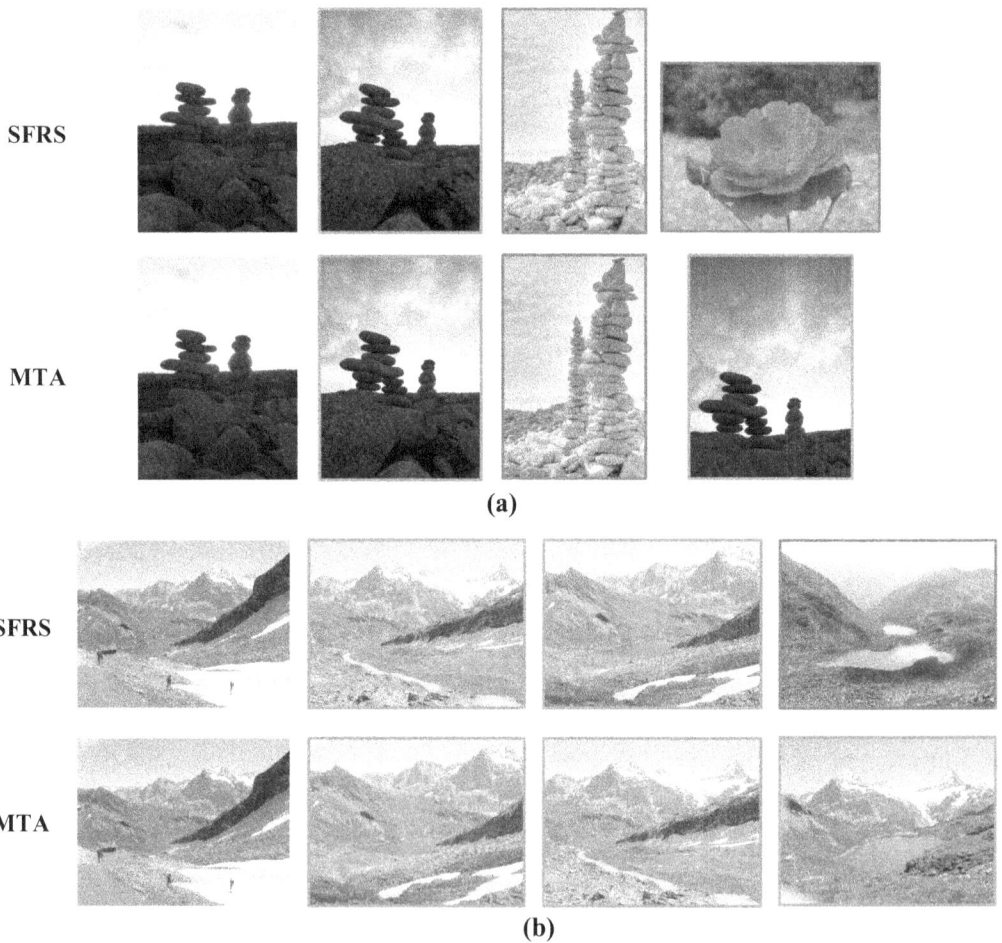

FIGURE 5.7 Visualization of the top-3 retrieved images of SFRS and our MTA for two selected query images from Holidays dataset. (a, b) The query image and the top-3 retrieved results, which are presented in the first column and last three columns, respectively.

FIGURE 5.8   Visualization of reference images provided by a visual sensor in an outdoor environment. (a) Building 1. (b) Building 2. (c) Building 3. (d) Building 4.

5.4.5 Evaluation on an Outdoor Environment

Figure 5.8 presents the reference images taken by a visual sensor toward four buildings with different positions and viewpoints on an outdoor environment shown in Figure 5.9. There are six images for each building. In Figure 5.9, four query images are collected at different viewpoints and illumination conditions from the reference images. We still adopt the global feature extraction network trained on Pitts30k dataset to extract the global features of reference images and query images.

By searching the nearest reference images in the feature space for each query image, we obtain the top-3 retrieval results of NetVLAD, SARE, SFRS, and our MTA, as illustrated in Figure 5.10. For the query image captured at night in Figure 5.10b, our method retrieves two correct reference images while other methods only find a matched result. All the methods retrieve the same top-1 reference image, which has a similar viewpoint with the query image.

FIGURE 5.9   Four selected query images at different positions of the outdoor environment.

The proposed method also retrieves another correct image with a large viewpoint difference from the query image. In Figure 5.10c, the top-3 retrieval results of our method are all correct, whereas other methods fail in the third retrieval result. Note that all the methods obtain the same top-2 retrieval images, and these images cover a large region of building 2. However, this is not the case for the third retrieval image of our method, where only a partial region of building 2 is included. The successful retrieval of this image demonstrates that our method can focus on the discriminative features.

5.4.6 Application in Relocalization of Indoor Robot Navigation

To further testify the proposed place recognition method MTA, we conduct robot navigation experiments in indoor corridor environment. In the robot navigation task, its 6 degrees of freedom (DOF) pose needs to be first obtained. Visual SLAM is a preferable solution to provide the robot pose. Due to the complexity of environment, the visual SLAM may fail. In this situation, relocalization based on the pre-built environment map becomes indispensable. In this experiment, the robot is normally localized through a visual SLAM method in Chapter 3. Once the failure of feature tracking occurs, the relocalization based on MTA is activated to recover the pose of robot. Herein, the pre-built map is built using the visual SLAM method, which contains multiple keyframes with feature points and corresponding 3D coordinates. By sparsely choosing images of keyframes, the reference images are acquired, whose global features are extracted offline using the trained global feature extraction network on Pitts30k dataset.

The detailed implementation of relocalization based on MTA is as follows. Given a query image, its global feature is extracted using MTA. Then, the reference images with small feature distances from the query image are mined. The smaller the Euclidean distance is, the better the reference image is. However, it is not accurate to directly use the pose of top-1 reference image as that of current query image because they are not completely the same. Besides, not every top-ranking reference image can be used to

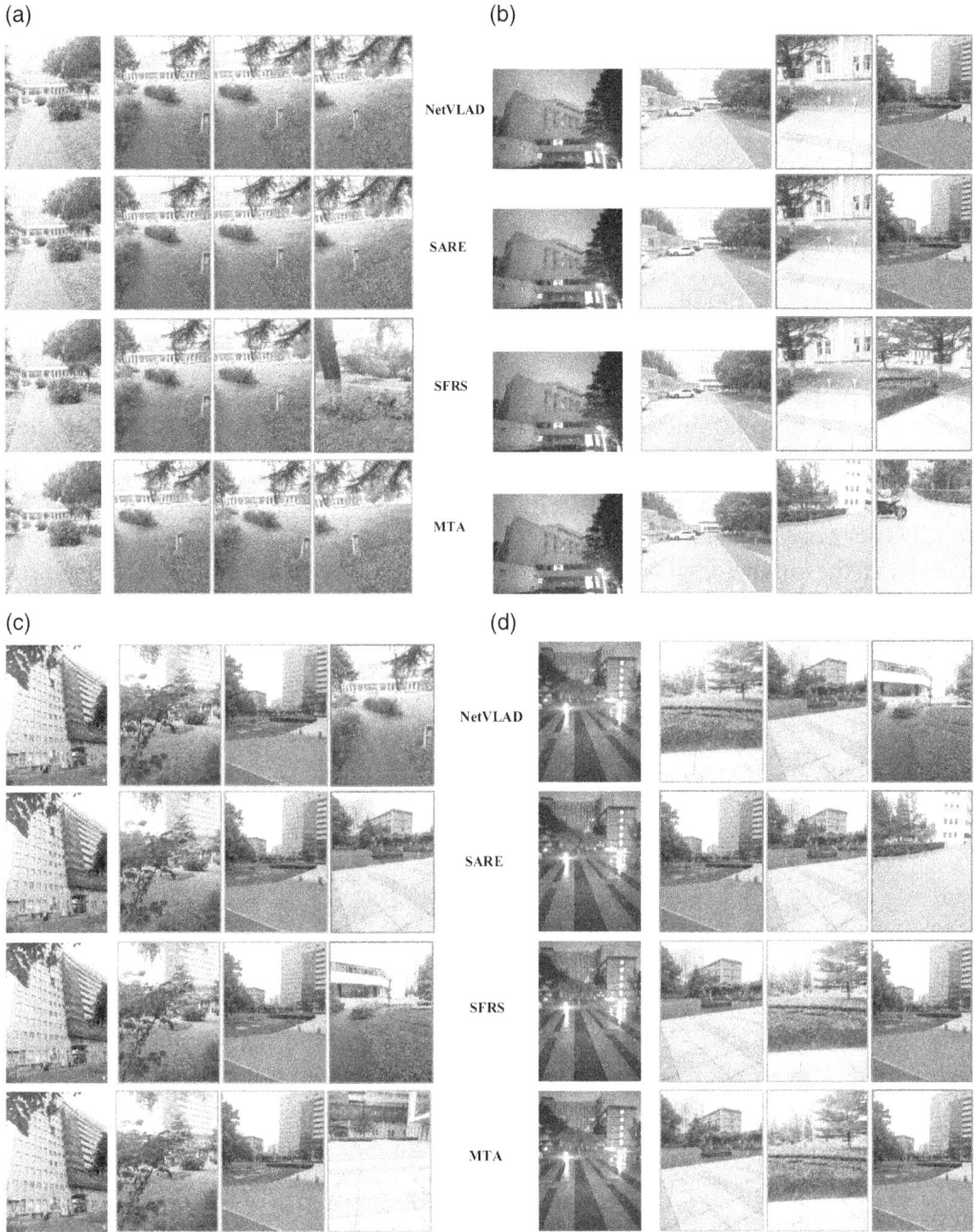

FIGURE 5.10   The top-3 retrieved images of different methods for four query images in Figure 5.9. (a)-(d) correspond to the images taken from different positions and viewpoints.

relocalize the query image accurately and only those with sufficient well-matched feature points are qualified. We denote the matched ORB feature point in query image and reference image with (q_i, p_i), $i = 1,2,\ldots,N_m$, $q_i = [u_i, v_i]^T$, where N_m refers to the number of matched feature points. If N_m is smaller than a given threshold N_{th}, this reference image is discarded. Let $P_i = \begin{bmatrix} x_i, y_i, z_i \end{bmatrix}^T$ represent the associated 3D coordinate of p_i in the pre-built environment map. According to the pinhole camera model in Chapter 2, one can obtain that

$$s_i \begin{bmatrix} u_p \\ v_p \\ 1 \end{bmatrix} = K \begin{bmatrix} R & t \end{bmatrix} \begin{bmatrix} x_i \\ y_i \\ z_i \\ 1 \end{bmatrix} \tag{5.19}$$

where $\begin{bmatrix} u_p, v_p \end{bmatrix}^T$ is the pixel coordinate of the projection of P_i on the query image, K is the camera intrinsic parameter, and R and t refer to the rotation matrix and translation vector of the query image, respectively. Herein, we resort to EPnP method [44] with random sample consensus (RANSAC) to solve the 6-DOF pose $\{R,t\}$ of the query image, which is further optimized following Chapter 3. We label $\{R^*, t^*\}$ as the optimized 6-DOF pose. Next, the number of good matched feature points is determined by calculating their respective re-projection errors:

$$error_i = \begin{bmatrix} \mathrm{err}_i^u \\ \mathrm{err}_i^v \end{bmatrix}^T \Sigma_i^{-1} \begin{bmatrix} \mathrm{err}_i^u \\ \mathrm{err}_i^v \end{bmatrix}$$

$$\begin{bmatrix} \mathrm{err}_i^u \\ \mathrm{err}_i^v \\ 0 \end{bmatrix} = \begin{bmatrix} u_i \\ v_i \\ 1 \end{bmatrix} - \frac{1}{s_i^*} K \begin{bmatrix} R^* & t^* \end{bmatrix} \begin{bmatrix} x_i \\ y_i \\ z_i \\ 1 \end{bmatrix} \tag{5.20}$$

where $error_i$ and Σ_i^{-1} stand for the re-projection error and information matrix of the ith feature point. When $error_i$ is smaller than a specific threshold err_{th}, the ith feature point is regarded as good. If the good feature points in a query image are enough, the pose $\{R^*, t^*\}$ is accepted as the relocalization result of the robot. Note that the above relocalization method is also employed in the first frame of the robot motion to obtain a global initial pose.

In the navigation experiment, the robot is required to move from the starting point S to the destination D based on a pre-built environment map with keyframes and map points, as shown in Figure 5.11. During the robot motion, an interference is exerted, where it is manually dragged approximately from location p_c to location p_d after its camera is totally occluded. When it is dragged to location p_d, the occlusion is removed, and then the relocalization succeeds. The first query image with a successful relocalization is illustrated in Figure 5.12a, and Figure 5.12a1–a3 presents the top-3 retrieved results of our MTA method. The numbers below are their feature distances to this query image. The video snapshots of

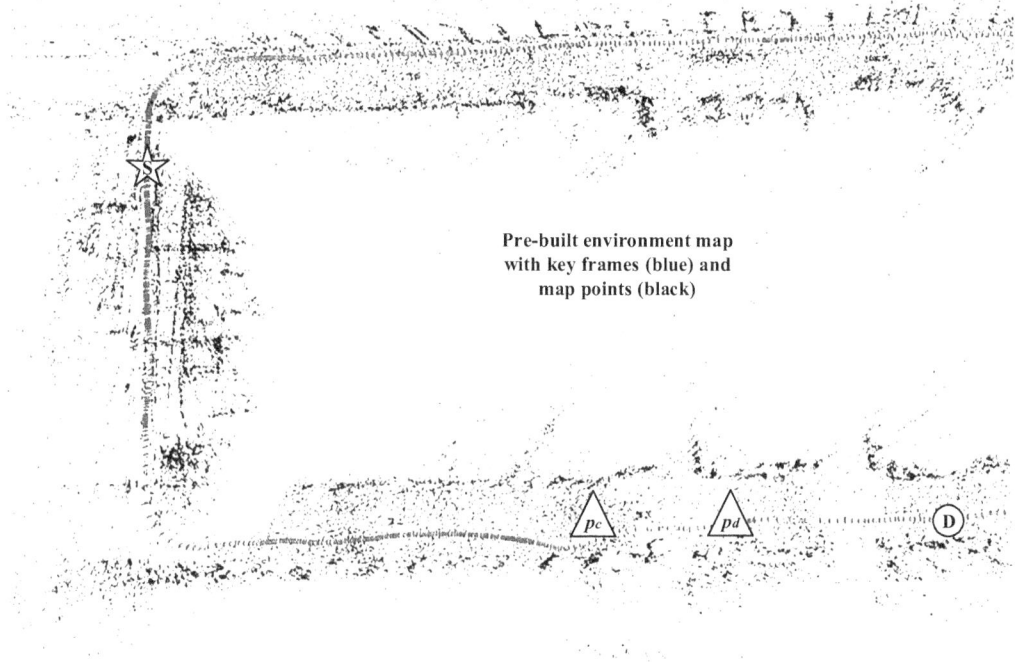

Pre-built environment map
with key frames (blue) and
map points (black)

FIGURE 5.11 Partial map of the experimental environment. The robot is manually dragged approximately from location p_c to location p_d, and the camera is totally occluded in this process. "S" and "D" represent the starting point and destination, respectively.

(a)　　　　　(a1) 1.025　　　　　(a2) 1.160　　　　　(a3) 1.407

FIGURE 5.12 The first query image with a successful relocalization at location p_d of Figure 5.11 as well as the top-3 retrieved results of our MTA method. The numbers below are their feature distances to the query image.

robot navigation are presented in Figure 5.13, where Figure 5.13e–g describes the dragging process. With a successful relocalization, the robot continues to move and finally arrives at its destination.

5.4.7 Discussion

The proposed place recognition method based on a multi-task learning framework provides an enhanced scheme to extract global features of images. The common triplet ranking loss imposes feature distance constraint only in triplets of a batch, which is actually

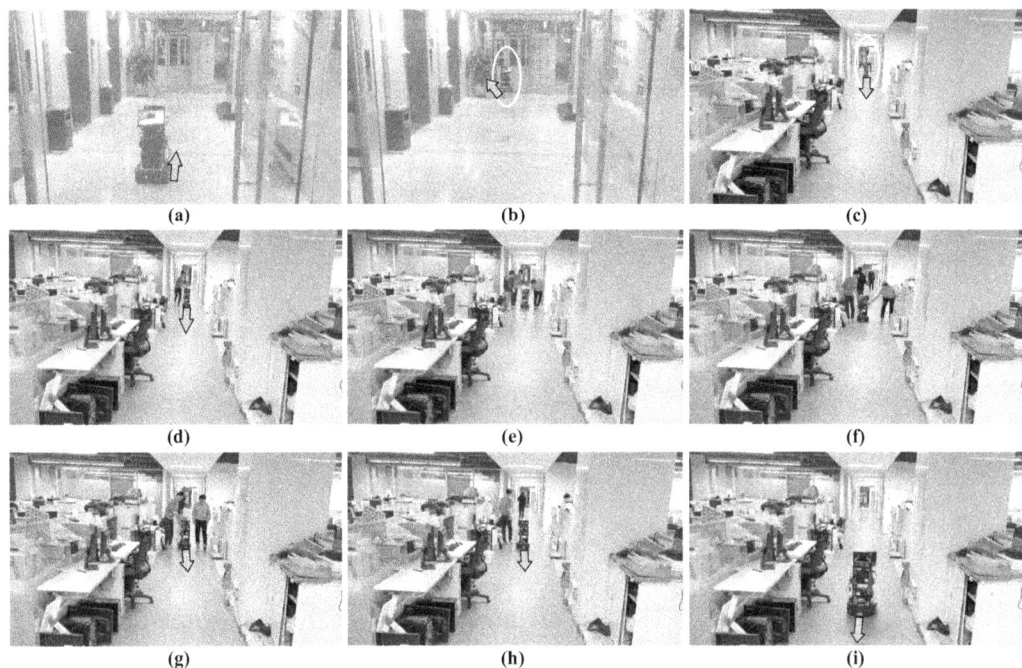

FIGURE 5.13 Video snapshots of robot navigation. Figure 5.13(d) corresponds to the time before camera occlusion. After the camera is totally occluded using a card, the robot is manually dragged from Figure 5.13(e)–(g).

an intra-batch constraint. On this basis, we introduce the binary classification loss and constrain the feature distances of all the positive pairs less than those of all the negative pairs, regardless of whether these pairs correspond to the same triplet/batch. It means that the binary classification loss considers the correlation of different batches. As a result, the global structure of the training dataset is exploited during the training, and the overfitting problem caused by the triplet ranking task is effectively prohibited. It is worth mentioning that the binary classification loss shall impair the order of similarity in appearance, which is compensated by combining the triplet ranking loss.

In related computer vision fields, the image retrieval problem is similar to the place recognition, and they endeavor to find the image similar to the query one from the reference images. However, there exists a difference between them. The image retrieval problem can utilize the class information of the training images, whereas the place recognition usually uses continuous GPS labels of each image, which makes the place recognition a weakly supervised problem with more challenges. Due to the weak supervision of GPS labels, two images with small geographical distances do not necessarily have overlapping regions. Many place recognition methods including our MTA have to select the most possible positive sample for each query image, according to the GPS information as well as the feature distance to the query image. A possible improvement is to take advantage of the information from multiple positive samples. The SFRS method [31] has made a beneficial attempt with good performance. SFRS utilizes a ranking loss with the most probable positive sample. Meanwhile, to further constrain the features of images, it also predicts the similarities between the query image and divided regions of other possible positive samples, which is trained in a self-supervised way. This image-to-region processing makes the network pay attention to the local information of images. In general, better mining of positive samples and self-supervised fine-grained image representation provide promising directions to further improve the network performance.

5.5 CONCLUSION

In this chapter, a multi-task learning framework with attention mechanism is proposed to enhance the performance of place recognition. A binary classification task including a binary classification network as well as a binary cross-entropy loss is designed to cooperate with the existing triplet ranking task. As a result, the generalization of our model is enhanced under the joint constraints of both tasks. Moreover, an attention module is presented to promote the model to pay more attention to salient regions for more effective global feature extraction. Experiment results on TokyoTM-val, Pitts250k-test, Tokyo 24/7, Holidays datasets, and actual environments verify the effectiveness of the proposed method.

REFERENCES

[1] Spera, E., Furnari, A., Battiato, S., & Farinella, G. M. (2019). EgoCart: A benchmark dataset for large-scale indoor image-based localization in retail stores. *IEEE Transactions on Circuits and Systems for Video Technology*, 31(4), 1253–1267.

[2] Lowe, D. G. (2004). Distinctive image features from scale-invariant keypoints. *International Journal of Computer Vision*, 60(2), 91–110.

[3] Philbin, J., Chum, O., Isard, M., Sivic, J., & Zisserman, A. (2007). Object retrieval with large vocabularies and fast spatial matching. In *Proceedings of the IEEE Conference on Computer Vision and Pattern Recognition*, Minneapolis, MN, USA (pp. 1–8).

[4] Arandjelovic, R., & Zisserman, A. (2013). All about VLAD. In *Proceedings of the IEEE Conference on Computer Vision and Pattern Recognition*, Portland, OR, USA (pp. 1578–1585).

[5] Xu, Y., Huang, J., Wang, J., Wang, Y., Qin, H., & Nan, K. (2021). ESA-VLAD: A lightweight network based on second-order attention and NetVLAD for loop closure detection. *IEEE Robotics and Automation Letters*, 6(4), 6545–6552.

[6] Zhang, X., Wang, L., & Su, Y. (2021). Visual place recognition: A survey from deep learning perspective. *Pattern Recognition*, 113, 107760.

[7] Zhai, Q., Huang, R., Cheng, H., Zhan, H., Li, J., & Liu, Z. (2020). Learning quintuplet loss for large-scale visual geolocalization. *IEEE MultiMedia*, 27(3), 34–43.

[8] Yu, J., Zhu, C., Zhang, J., Huang, Q., & Tao, D. (2019). Spatial pyramid-enhanced NetVLAD with weighted triplet loss for place recognition. *IEEE Transactions on Neural Networks and Learning Systems*, 31(2), 661–674.

[9] Liu, L., Li, H., & Dai, Y. (2019). Stochastic attraction-repulsion embedding for large scale image localization. In *Proceedings of the IEEE International Conference on Computer Vision*, Seoul, Korea (South) (pp. 2570–2579).

[10] Thoma, J., Paudel, D. P., Chhatkuli, A., & Van Gool, L. (2020). Geometrically mappable image features. *IEEE Robotics and Automation Letters*, 5(2), 2062–2069.

[11] Thoma, J., Paudel, D. P., Chhatkuli, A., & Van Gool, L. (2020). Learning condition invariant features for retrieval-based localization from 1M images. arXiv preprint arXiv:2008.12165.

[12] Thoma, J., Paudel, D. P., & Gool, L. V. (2020). Soft contrastive learning for visual localization. *Advances in Neural Information Processing Systems*, 33, 11119–11130.

[13] Leyva-Vallina, M., Strisciuglio, N., & Petkov, N. (2021). Generalized contrastive optimization of siamese networks for place recognition. arXiv preprint arXiv:2103.06638.

[14] Kim, H. J., Dunn, E., & Frahm, J. (2017). Learned contextual feature reweighting for image geo-localization. In *Proceedings of the IEEE Conference on Computer Vision and Pattern Recognition*, Honolulu, HI, USA (pp. 3251–3260).

[15] Chen, Z., Liu, L., Sa, I., Ge, Z., & Chli, M. (2018). Learning context flexible attention model for long-term visual place recognition. *IEEE Robotics and Automation Letters*, 3(4), 4015–4022.

[16] Xin, Z., Cai, Y., Lu, T., Xing, X., Cai, S., Zhang, J., Yang, Y., & Wang, Y. (2019). Localizing discriminative visual landmarks for place recognition. In *Proceedings of the International Conference on Robotics and Automation*, Montreal, QC, Canada (pp. 5979–5985).

[17] Chen, W., Chen, X., Zhang, J., & Huang, K. (2017). A multi-task deep network for person re-identification. *Proceedings of the AAAI Conference on Artificial Intelligence*, 31(1), 3988–3994.

[18] Wen, Y., Zhang, K., Li, Z., & Qiao, Y. (2016). A discriminative feature learning approach for deep face recognition. In *Proceedings of the European Conference on Computer Vision*, Amsterdam, The Netherlands (pp. 499–515).

[19] Arandjelovic, R., Gronat, P., Torii, A., Pajdla, T., & Sivic, J. (2016). NetVLAD: CNN architecture for weakly supervised place recognition. In *Proceedings of the IEEE Conference on Computer Vision and Pattern Recognition*, Las Vegas, NV, USA (pp. 5297–5307).

[20] Bay, H., Tuytelaars, T., & Van Gool, L. (2006). SURF: Speeded up robust features. In *Proceedings of the European Conference on Computer Vision*, Graz, Austria (pp. 404–417).

[21] Calonder, M., Lepetit, V., Strecha, C., & Fua, P. (2010). BRIEF: Binary robust independent elementary features. In *Proceedings of the European Conference on Computer Vision*, Heraklion, Crete, Greece (pp. 778–792).

[22] Leutenegger, S., Chli, M., & Siegwart, R. Y. (2011). BRISK: Binary robust invariant scalable keypoints. In *Proceedings of the International Conference on Computer Vision*, Barcelona, Spain (pp. 2548–2555).

[23] Perronnin, F., & Dance, C. (2007). Fisher kernels on visual vocabularies for image categorization. In *Proceedings of the IEEE Conference on Computer Vision and Pattern Recognition*, Minneapolis, MN, USA (pp. 1–8).

[24] Dalal, N., & Triggs, B. (2005). Histograms of oriented gradients for human detection. In *Proceedings of the IEEE Conference on Computer Vision and Pattern Recognition*, San Diego, CA, USA (pp. 886–893).

[25] Cummins, M., & Newman, P. (2008). FAB-MAP: Probabilistic localization and mapping in the space of appearance. *The International Journal of Robotics Research*, 27(6), 647–665.

[26] Gálvez-López, D., & Tardos, J. D. (2012). Bags of binary words for fast place recognition in image sequences. *IEEE Transactions on Robotics*, 28(5), 1188–1197.

[27] Mur-Artal, R., & Tardós, J. D. (2014). Fast relocalisation and loop closing in keyframe-based SLAM. In *Proceedings of the IEEE International Conference on Robotics and Automation*, Hong Kong, China (pp. 846–853).

[28] Jie, Z., Lu, W. F., Sakhavi, S., Wei, Y., Tay, E. H. F., & Yan, S. (2016). Object proposal generation with fully convolutional networks. *IEEE Transactions on Circuits and Systems for Video Technology*, 28(1), 62–75.

[29] Sun, T., Liu, M., Ye, H., & Yeung, D. Y. (2019). Point-cloud-based place recognition using CNN feature extraction. *IEEE Sensors Journal*, 19(24), 12175–12186.

[30] Sünderhauf, N., Shirazi, S., Jacobson, A., Dayoub, F., Pepperell, E., Upcroft, B., & Milford, M. (2015). Place recognition with convnet landmarks: Viewpoint-robust, condition-robust, training-free. In *Robotics: Science and Systems*, Rome, Italy (pp. 1–10).

[31] Ge, Y., Wang, H., Zhu, F., Zhao, R., & Li, H. (2020). Self-supervising fine-grained region similarities for large-scale image localization. In *Proceedings of the European Conference of Computer Vision*, Glasgow, UK (pp. 369–386).

[32] Deng, J., Dong, W., Socher, R., Li, L., Li, K., & Li, F. (2009). ImageNet: A large-scale hierarchical image database. In *Proceedings of the IEEE Conference on Computer Vision and Pattern Recognition*, Miami, FL, USA (pp. 248–255).

[33] Yu, F., Koltun, V., & Funkhouser, T. (2017). Dilated residual networks. In *Proceedings of the IEEE Conference on Computer Vision and Pattern Recognition*, Honolulu, HI, USA (pp. 472–480).

[34] Zhang, F., Qi, X., Yang, R., Prisacariu, V., Wah, B., & Torr, P. (2020). Domain-invariant stereo matching networks. In *Proceedings of the European Conference on Computer Vision*, Glasgow, UK (pp. 420–439).

[35] Lagunes-Fortiz, M., Damen, D., & Mayol-Cuevas, W. (2019). Learning discriminative embeddings for object recognition on-the-fly. In *Proceedings of the IEEE International Conference on Robotics and Automation*, Montreal, QC, Canada (pp. 2932–2938).

[36] Kumar, B. G. V., Carneiro, G., & Reid, I. (2016). Learning local image descriptors with deep siamese and triplet convolutional networks by minimising global loss functions. In *Proceedings of the IEEE Conference on Computer Vision and Pattern Recognition*, Las Vegas, NV, USA (pp. 5385–5394).

[37] Torii, A., Sivic, J., Pajdla, T., & Okutomi, M. (2013). Visual place recognition with repetitive structures. In *Proceedings of the IEEE Conference on Computer Vision and Pattern Recognition*, Portland, OR, USA (pp. 883–890).

[38] Torii, A., Arandjelovic, R., Sivic, J., Okutomi, M., & Pajdla, T. (2015). 24/7 place recognition by view synthesis. In *Proceedings of the IEEE Conference on Computer Vision and Pattern Recognition*, Boston, MA, USA (pp. 1808–1817).

[39] Revaud, J., Almazán, J., Rezende, R. S., & Souza, C. R. D. (2019). Learning with average precision: Training image retrieval with a listwise loss. In *Proceedings of the IEEE International Conference on Computer Vision*, Seoul, Korea (South) (pp. 5107–5116).

[40] conv5-triplet-lr0.001-tuple4. https://drive.google.com/drive/folders/1sjsxnlqyUCjrXOg2ahwr f7WufU1jvO_D?usp=sharing.

[41] Model Zoo. https://github.com/yxgeee/OpenIBL/blob/master/docs/MODEL_ZOO.md.

[42] Zhu, Y., Wang, J., Xie, L., & Zheng, L. (2018). Attention-based pyramid aggregation network for visual place recognition. In *Proceedings of the ACM International Conference on Multimedia*, Seoul Republic of Korea (pp. 99–107).

[43] Jegou, H., Douze, M., & Schmid, C. (2008). Hamming embedding and weak geometric consistency for large scale image search. In *Proceedings of the European Conference on Computer Vision*, Marseille, France (pp. 304–317).

[44] Lepetit, V., Moreno-Noguer, F., & Fua, P. (2009). EPnP: An accurate O(n) solution to the PnP problem. *International Journal of Computer Vision*, 81(2), 155–166.

CHAPTER 6

Robot Visual Localization Framework Based on Offline Hybrid Map

6.1 INTRODUCTION

The PO-SLAM method is suitable for real-time relative pose estimation and mapping during robot motion; the SFT-CR relocalization method is applicable to small-scale scenes with high localization accuracy requirements and often more robust to conditions such as low texture and environmental variations, where the scene structure information is embedded in the trained scene coordinate regression network; while MTA is primarily used for larger-scale environments with relatively lower localization requirements, where the global features of the image can be used to determine the similarity between the query image and reference images in the map, narrowing the matching range and improving the relocalization efficiency. As mentioned earlier, SLAM tends to drift over time during extended motion. Large movements or environments with fewer features may cause tracking failure, and relying solely on an online map makes it difficult to recover the pose. Additionally, the localization result of SLAM is the pose of the current frame relative to the starting frame, while actual tasks typically rely on global pose information. To mitigate the above challenges, constructing an offline global map in advance and executing global relocalization based on this map to complement SLAM offers an effective solution. On the other hand, the robot's actual operating environment is complex, necessitating different relocalization strategies for various scenarios. In certain areas (such as corridors), the robot focuses more on motion than on precise localization. In other critical areas (such as designated workspaces or rooms), the robot needs precise localization while moving to perform tasks better. Therefore, it is necessary to build corresponding offline submaps for different relocalization methods, which are determined according to the distinct characteristics of different regions. Herein, SFT-CR and MTA methods respectively correspond to an implicit map represented with scene coordinate regression (SCoRe) network and an

DOI: 10.1201/9781003643630-6

explicit map with keyframes and map points. They can be combined to form an offline hybrid map. Based on this, continuous and stable real-time global localization of the robot can be achieved by relying on the relocalization functionality of both SFT-CR and MTA methods, combined with the PO-SLAM method.

In this chapter, we propose a visual localization software architecture for service robots based on an offline hybrid map. According to the size of the environment and the task requirement, the hybrid map combining explicit and implicit submaps is constructed based on the adaptability of explicit map-based relocalization method to large-scale environments and the relocalization accuracy of SCoRe network-based method in small scenes. Specifically, the explicit submap of the environment is built by combining PO-SLAM and the global feature extraction network of MTA. The implicit submap is stored in the scene coordinate regression network SFT-CR. In addition, a reference coordinate system is also defined to align the submaps of different regions. On this basis, PO-SLAM is combined with MTA and SFT-CR methods to achieve stable robot localization. When PO-SLAM tracking fails, the robot recovers its pose via either MTA or SFT-CR, according to the submap where the robot is located. The contributions of our method are summarized as follows:

1. We propose a visual localization soft architecture based on an explicit and implicit combined offline hybrid map, which achieves fast and stable global localization for robots.

2. An offline hybrid map construction module is designed, where the hybrid map of environment is divided into multiple explicit and implicit submaps based on the size and task requirements of different regions within the environment. These submaps are further merged and uniformed to a reference coordinate system for global localization.

3. We present a submap selection strategy to recover the robot pose in case of SLAM tracking failure. It gives priority to the submap of the region where the robot was previously located. Furthermore, the system quickly filters out unsuitable explicit submaps based on the global feature similarity between the current image and each explicit submap. Navigation experiments conducted in an indoor office environment demonstrate that the proposed localization framework can achieve continuous and stable localization for robots in large-scale environments.

The rest of this chapter is structured as follows. Section 6.2 presents the related work. The proposed methodology is described in Section 6.3 in detail. Experimental results are given in Section 6.4, and Section 6.5 concludes this chapter.

6.2 REVIEW OF EXPLICIT MAP-BASED AND IMPLICIT MAP-BASED RELOCALIZATION

For explicit map-based relocalization methods, explicit 2D–3D matching is essential. Sattler et al. [1] proposed a direct matching framework based on visual vocabulary quantization and a prioritized correspondence search. This method suffers from quantization artifacts:

if image features and their corresponding 3D points are assigned to different words, correct matches become difficult to find. To address this, Sattler et al. [2] introduced an active search mechanism. They first match the 2D feature points in the query image to the 3D points in the map, and then search for neighbors of these matched 3D points and actively match these neighbors to the features in the query image, reducing the impact of quantization artifacts. However, these methods all rely on traditional hand-crafted feature points, which mainly focus on low-level texture structures and lack robustness against environmental changes in long-term localization tasks [3]. The rise of deep learning has led to the development of neural network-based feature extraction and description methods. Yi et al. [4] proposed a novel deep network architecture, which learns to handle keypoint detection, orientation estimation, and feature description tasks in a unified way while maintaining end-to-end differentiability. For feature matching, Sarlin et al. [5] introduced SuperGlue, which uses an attention-based graph neural network to enhance the representation of feature points, and an optimal matching layer to compute a match score matrix among feature points of two images and solves for the optimal feature assignment matrix using the Sinkhorn algorithm [6]. On the other hand, the introduction of visual place recognition [7,8] increases the scalability of localization to large-scale environments. However, above methods still rely on local feature matching, and it can also impact localization accuracy when the number of correctly matched features is low.

As mentioned in Chapter 1, implicit map-based visual localization methods include global pose regression-based method and scene coordinate regression-based method. For global pose regression-based methods, PoseNet [9] is the most representative method, which exploits a CNN pre-trained by classification task on giant datasets to solve the camera pose. Afterwards, some improvements arise including introducing Bayesian CNN for the estimation of relocalization uncertainty [10], incorporating geometric loss [11], exploiting the constraint of temporal smoothness [12], and integrating self-attention mechanism [13]. Nevertheless, such method estimates a relatively coarse camera pose since it does not make full use of the 3D scene structure and is naturally similar to image retrieval-based localization [14]. Compared to the above explicit map-based and global pose regression-based methods, scene coordinate regression-based approaches generally offer higher localization accuracy but are applicable to small-scaled scenes.

6.3 VISUAL LOCALIZATION SYSTEM BASED ON EXPLICIT AND IMPLICIT COMBINED HYBRID MAP

The framework of the proposed visual localization method based on an explicit and implicit combined offline hybrid map is illustrated in Figure 6.1, which consists of two components: offline hybrid map construction and online robot localization based on the offline map. The online localization part relies on SLAM tracking for fast relative pose estimation, which is combined with the explicit/implicit submap-based relocalization to address the cases of global pose estimation of the first frame and tracking failure. As a result, a stable robot global localization architecture is attained.

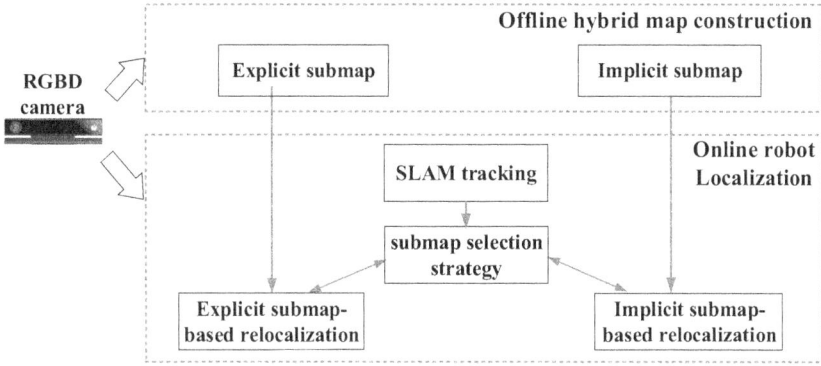

FIGURE 6.1   The framework of the proposed visual localization architecture based on an explicit and implicit combined offline hybrid map.

6.3.1 Offline Hybrid Map Construction

6.3.1.1 Explicit Map Construction

Here, we combine our SLAM method PO-SLAM with the global feature extraction network of MTA to construct the explicit map for a given environment. As a result, the constructed offline map includes not only the keyframe RGB images, poses, ORB features, corresponding map points, and co-visibility relationships between keyframes obtained through PO-SLAM, but also the global features provided by MTA. These global features are extracted by inputting the above keyframe RGB images into the global feature extraction network that was trained on the Pitts30k dataset, resulting in features with a dimension of 4,096. To accelerate the subsequent relocalization, the global features of the keyframes are sparsely sampled as the reference global features. Considering the significant changes in camera perspectives at corners, a reference image is selected every 5 keyframes in these areas. In regions with smaller changes in perspective, a reference image is selected every 10 keyframes.

6.3.1.2 Implicit Map Construction

The implicit map is stored in the form of model parameters, and its construction requires training the corresponding model. Here, we use the scene coordinate regression network with spatial feature transformation (SFT-CR). Training this network requires RGB images and their per-pixel 3D coordinate labels as training data. To achieve this, we first utilize a SfM method to process the input RGB and depth images and compute the corresponding 6D poses $\mathbf{T}_{wc}$ of the RGB images in the coordinate system whose origin is defined by the first frame of the camera. By combining the depth images with the camera projection model, we can obtain the 3D world coordinates corresponding to each pixel of the RGB images.

Suppose each pixel coordinate in the RGB image is $\mathbf{p} = [u, v]^T$, and the coordinate of the 3D point corresponding to pixel $\mathbf{p}$ in the current camera coordinate system is $\mathbf{P}_c = [X_c, Y_c, Z_c]^T$, where Z_c is the depth value corresponding to pixel $\mathbf{p}$. According to the pinhole camera model in Chapter 2, we can obtain

$$Z_c \begin{bmatrix} u \\ v \\ 1 \end{bmatrix} = \begin{bmatrix} f_x & 0 & c_x \\ 0 & f_y & c_y \\ 0 & 0 & 1 \end{bmatrix} \begin{bmatrix} X_c \\ Y_c \\ Z_c \end{bmatrix} \tag{6.1}$$

where f_x and f_y are the focal lengths of the camera along the x and y axes, respectively, and (c_x, c_y) are the pixel coordinates of the principal point. Therefore, we have

$$X_c = \frac{(u - c_x) Z_c}{f_x}$$
$$Y_c = \frac{(v - c_y) Z_c}{f_y} \tag{6.2}$$

Therefore, the 3D coordinate $\mathbf{P}_w$ of the map point corresponding to pixel p in the coordinate system of the first camera frame is

$$\mathbf{P}_w = \mathbf{T}_{wc} \mathbf{P}_c \tag{6.3}$$

Based on the above RGB images and the obtained dense 3D coordinates, we train the scene coordinate regression network SFT-CR to achieve an implicit scene representation.

6.3.1.3 Submaps Merging

When constructing a hybrid map for a large-scale environment, the submaps of different regions need to be unified into a common coordinate system. Typically, an explicit map is selected as the reference map, with its first-frame coordinate system serving as the reference coordinate system. Then, we compute the poses of the first frame of other submaps in the reference coordinate system to obtain the transformation matrices between other submaps and the reference map. Let the reference coordinate system be W, and the first-frame coordinate system of certain submap be w. To compute the transformation matrix $\mathbf{T}_{Ww}$ from coordinate system w to W, the robot first moves from the reference map W with PO-SLAM continuously tracking and localizing. The robot stops at the vicinity of the position corresponding to the first frame of submap w, so that there is large viewpoint overlapping between these two positions. At the robot's stopping position, the global pose of the image, ORB feature points, and the corresponding 3D coordinates of the feature points are obtained by PO-SLAM. We perform feature matching between this image and the stored first-frame image of submap w, and the matched 2D–3D point pairs are input into a PnP solver [15] to compute the transformation matrix $\mathbf{T}_{Ww}$. Thus, each submap w can be aligned to reference map W to form a global hybrid map.

Algorithm 6.1 presents the construction algorithm for the robot's offline hybrid map, where PO-SLAM(·) denotes the PO-SLAM localization and mapping process, trainedSFT-CR(·) represents the training process of the SFT-CR network, Num(·) indicates the number of elements in an array, UnProj(·) is the operation of back-projecting pixels in an image to the 3D space, Transform(·) represents the computation of transformation matrix from the coordinate system of the input submap to the reference coordinate system, Sample(·) is used to sample the global features of keyframes in the explicit submap to obtain the global features of reference images in this submap.

Algorithm 6.1 Robot Offline Hybrid Map Construction

Input: Number M of explicit submaps and the image sequences ES_i required for their construction, $i = 1, 2, \ldots, M$, number N of implicit submaps and the image sequences IS_k required for their construction $k = 1, 2, \ldots, N$, the network MTA_model() trained on the Pitts30k dataset.

Output: The set of explicit submaps ExpMaps and their transformation matrices to the reference coordinate system TransExpMaps, and the set of implicit submaps ImMaps and their transformation matrices to the reference coordinate system TransImMaps.

1. **for** $i = 1, 2, \ldots, M$
2.   {KFs, KFs_orbs, KFs_points, KFs_connections}=PO-SLAM(ES_i);
3.   ExpMaps[i].push_back({KFs, KFs_orbs, KFs_points, KFs_connections});
4.   **for** $j = 1, 2, \ldots$, Num(KFs)
5.   KFs_glof.push_back(MTA_model(KFs[j]);
6.   **end for**
7.   RFs_glof=Sample(KFs_glof);
8.   ExpMaps[i].push_back(RFs_glof);
9.   TransExpMaps[i]=Transform(ExpMaps[i]);
10. **end for**
11. **for** $k = 1, \ldots, N$
12.   {TrainedRGBs, TrainedDepths, TrainedPoses}=SFM(IS_k);
13.   **for** $j = 1, 2, \ldots$, Num(TrainedRGBs)
14.   CoordMaps.push_back(UnProj(TrainedRGBs[j], TrainedDepths[j], TrainedPoses[j]));
15.   **end for**
16.   ImMaps.push_back(TrainedSFT-CR(TrainedRGBs, CoordMaps));
17.   TransImMaps[k]=Transform(ImMaps[k]);
18. **end for**
19. **return** ExpMaps, TransExpMaps, ImMaps, TransImMaps

6.3.2 Online Robot Localization Based on Offline Hybrid Map

The online localization part combines PO-SLAM's tracking and relocalization based on explicit or implicit submap to continuously provide the robot with global 6D pose, where relocalization methods MTA and SFT-CR are aimed at explicit and implicit submaps, respectively. During the robot's movement, SLAM is used for fast and continuous relative pose estimation. Relocalization is executed when the robot begins to move to obtain the initial global pose. Besides, it is also activated when PO-SLAM tracking fails to restore the global pose.

Specially, a submap selection strategy is designed to determine the sub-region of the robot. On the basis, relocalization based on the selected submap is executed to obtain the global pose.

6.3.2.1 Submap Selection Strategy

During relocalization, the robot first selects the corresponding submap based on its previous position and performs the explicit/implicit submap-based relocalization. If the relocalization is successful, the camera pose is restored, and PO-SLAM tracking continues. Otherwise, the robot attempts relocalization using submaps from other regions, where explicit submaps are prioritized for evaluation according to global feature similarity to accelerate the search process. Concretely, the global feature extraction network from MTA is first used to extract the global feature of the current image f_q. Then the feature distance $f_{qe_{ij}}$ between the current image and jth reference image in the ith explicit submap is calculated, and the inverse of the smallest feature distance is used as the global feature similarity f_{qe_i} between the current image and the explicit submap as follows:

$$f_{qe_i} = \frac{1}{\min\left\{ f_{qe_{ij}} \big\|_{j=1}^{N_i} \right\}} = \frac{1}{\min\left\{ \left\| f_q - f_{e_{ij}} \right\| \big\|_{j=1}^{N_i} \right\}} \tag{6.4}$$

where N_i and $f_{e_{ij}}$ represent the number of reference images and global feature of jth reference image in the ith explicit submap, respectively. Explicit submaps with low global feature similarities to the current image are excluded from consideration. Those that pass the similarity test are prioritized for relocalization, following the procedure outlined in Chapter 5, in descending order of similarity. If relocalization fails after evaluating these explicit submaps, implicit submaps are then considered for relocalization, as described in Chapter 4. Should relocalization still be unsuccessful after evaluating all regions, the robot proceeds to repeat the process for the next frame. Once relocalization succeeds, the current image's pose $\mathbf{T}_{Wc}$ in the reference coordinate system is obtained, thereby allowing the robot's movement to be resumed.

The localization algorithm based on the offline hybrid map is shown in Algorithm 6.2. In this algorithm, MTA(·) and SFT-CR(·) perform MTA-based relocalization and SFT-CR-based relocalization, respectively, using the input RGB image and its corresponding submap. GlobalPose(·) outputs the pose of the current image in the reference coordinate system, ComputeSim(·) calculates the global feature similarity between the input image and each submap, and CurZone(·) determines the robot's current region based on the input pose information.

Algorithm 6.2 Localization Based on Offline Hybrid Map

Input: Input image (Col, Dep), the set ExpMaps of explicit maps and their transformation matrices TransExpMaps to the reference coordinate system, the set ImMaps of implicit maps and their transformation matrices TransImMaps to the reference coordinate system, the global feature extraction network MTA_model() trained on the Pitts30k dataset.

Output: the global pose.

1. flag=False, Place=−1;
2. **if** flag is True
3. T = PO-SLAM(Col, Dep);
4. **else**
5.   **if** Place >= 1 ∩ Place <= M
6. T = MTA(Place, Col);
7.   **if** T is not Null
8. T = GlobalPose(T, TransExpMaps[Place]);
9.   **end if**
10. **end if**
11. **if** Place >= M + 1 ∩ Place <= M + N
12. T = SFT-CR(Place-M, Col);
13.   **if** T is not Null
14. T = GlobalPose(T, TransImMaps[Place-M]);
15.   **end if**
16. **end if**
17. **if** Place=−1 ∪ T is Null
18.   **for** i = 1, 2, …, M
19.   **if** ComputeSim(MTA_model(Col), ExpMaps[i])> Sim_{th}
20. T = MTA(i, Col);
21.   **if** T is not Null
22. T = GlobalPose(T, TransExpMaps[i]); break;
23.   **end if**
24.   **end if**
25.   **end for**
26.   **if** T is Null
27.   **for** k = 1, 2, …, N
28. T = SFT-CR(k, Col);
29.   **if** T is not Null
30. T = GlobalPose(T, TransImMaps[k]); break;
31.   **end if**
32.   **end for**
33.   **end if**
34. **end if**
35. **end if**
36. **if** T is not Null
37.   Place=CurZone(T); flag=True;
38. **else**
39.   flag=False;
40. **end if**
41. **return** T

6.4 EXPERIMENTS

To verify the effectiveness of the proposed service robot localization system, navigation experiments were conducted in an indoor office environment. We adopt the EAIBOT SSB1 mobile platform, with Kinect v2 as the visual sensor, a Velodyne as the LiDAR sensor, and a ZOTAC control processor equipped with an Intel i7–7700 CPU and an NVIDIA GeForce GTX 1080 GPU, running on Ubuntu 16.04 operating system. Note that the visual and LiDAR sensors are used for localization and obstacle avoidance, respectively. The point cloud data collected by the LiDAR is employed to construct the occupancy grid map for avoiding the local obstacles. Combining with the Vector Field Histogram (VFH) algorithm [16] and the online localization information, the robot's navigation control is achieved.

Figure 6.2 shows the indoor environment used for the experiment, including corridor and room areas. The line represents the robot's trajectory while constructing the offline explicit submap of the corridor environment. Additionally, an implicit submap using the scene coordinate regression network was built for the important area, Room 03.

6.4.1 Localization and Navigation Experiments in a Corridor Environment

In Experiment 1, the robot was tasked with navigating from the entrance of Room 03 to Elevator 2. The corresponding offline explicit map is shown in Figure 6.3, where the cones represent the keyframes in the map, the dots represent map points, and the lines between the keyframes represent the co-visibility relationships between them. "S" denotes the starting point, and "D" represents the destination. Additionally, Figure 6.3 provides

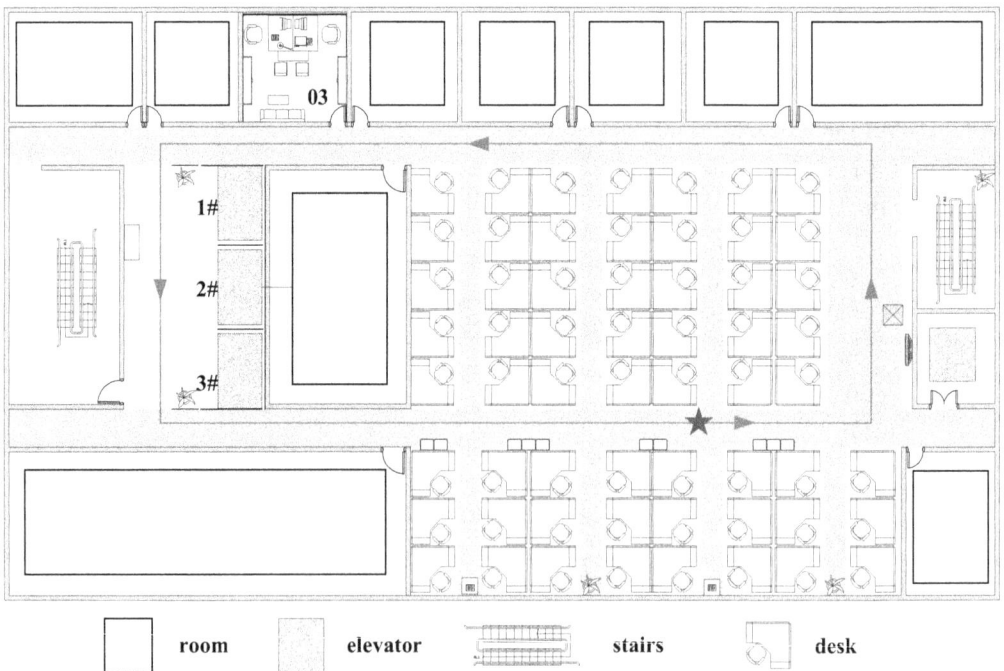

FIGURE 6.2   The experimental environment.

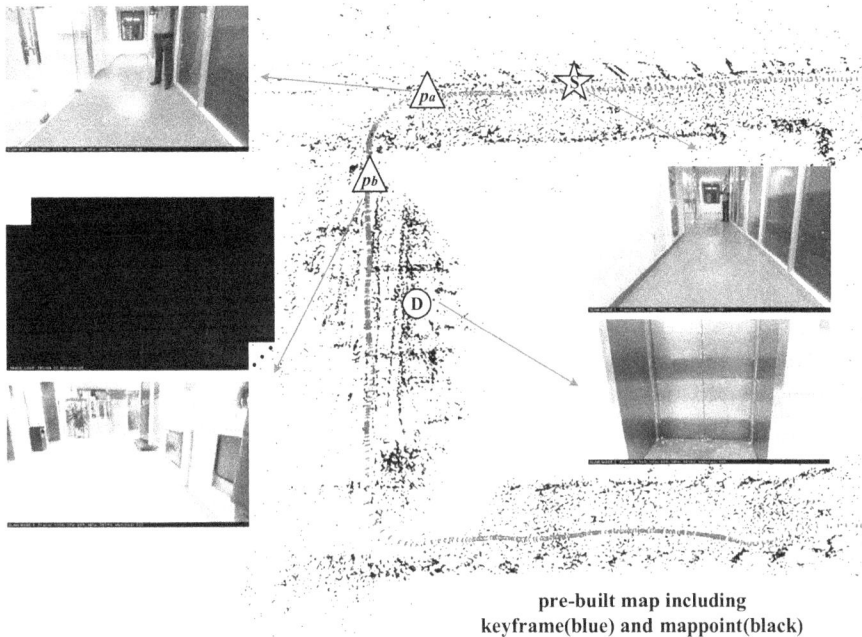

pre-built map including keyframe(blue) and mappoint(black)

FIGURE 6.3 The offline explicit map along with grayscale images from the robot's perspective for Experiment 1. Each grayscale image highlights point and object features. At position p_b, the camera was fully occluded for approximately ten seconds.

grayscale images from the robot's perspective at four different locations, along with their corresponding point and object semantic features. At the beginning, the robot loaded the offline-constructed map and performed relocalization. The query image and the top-3 retrieved reference images during relocalization are shown in Figure 6.4a and a1–a3, respectively. The numbers below the images represent the feature distances between each reference image and the query image in the first column. The robot uses the retrieved reference image for relocalization to obtain the pose of the first frame in the reference coordinate system. During the robot's movement, the camera is intentionally covered for approximately 10 seconds at position p_b. During this time, the camera could only capture black images, causing relocalization to fail. As a result, the robot stops at p_b since it could not obtain its current location. After the obstruction is removed, the robot successfully relocalizes using the query image captured at p_b (see Figure 6.4b). Despite the significant distortion caused by the sudden transition from darkness to brightness, the method still successfully retrieves reference images near the query image's location, as shown in Figure 6.4b1–b3. Video screenshots of the robot's navigation during Experiment 1 are shown in Figure 6.5, where Figure 6.5d–f corresponds to the moments before, during, and after the obstruction. The robot ultimately successfully reached and faced Elevator 2.

Experiment 2 focuses on relocalization in a corner area. After successfully re-localizing at point S, as shown in Figure 6.6, the robot begins moving. At position p_a, the camera is fully obstructed for approximately 10 seconds, during which the robot stops at p_a.

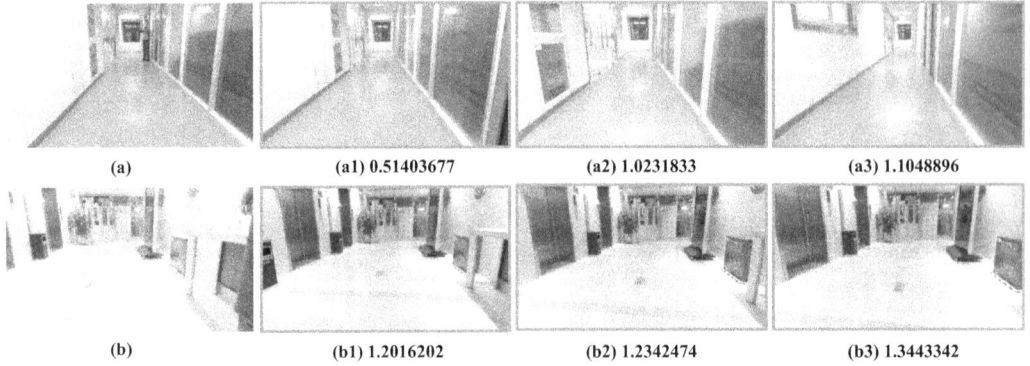

FIGURE 6.4   Place recognition results for relocalization in Experiment 1. (a) and (a1)–(a3) show the successfully relocalized query image at the starting point S and its top-3 retrieval results, while (b) and (b1)–(b3) show the successfully relocalized query image at location p_b and its top-3 retrieval results. The numbers below each image represent the feature distance between that image and the query image.

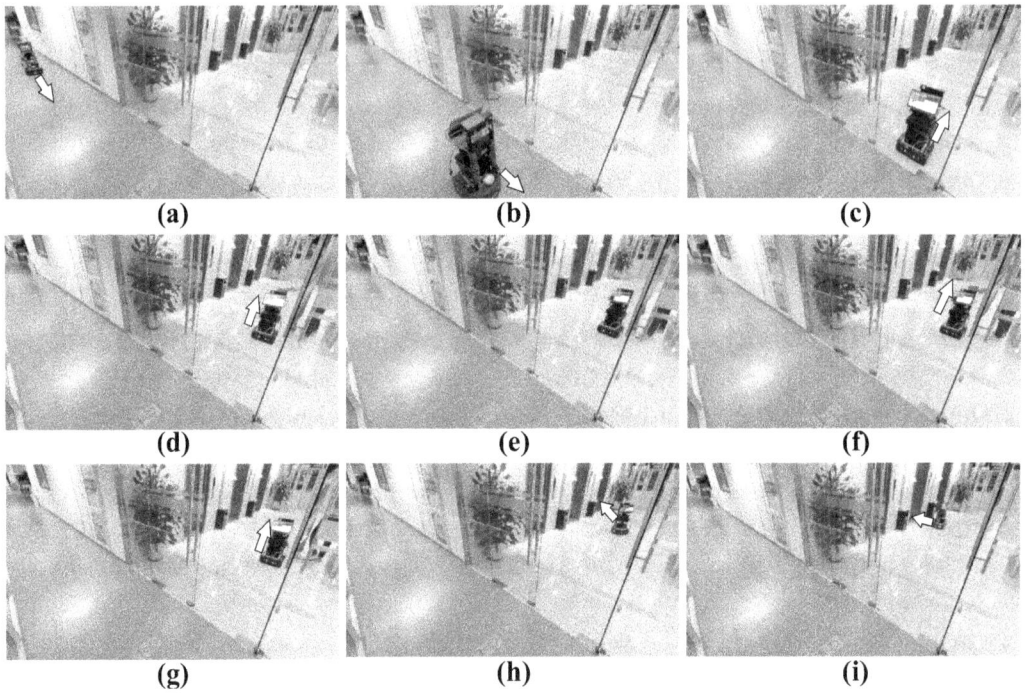

FIGURE 6.5   Video snapshots of robot navigation in Experiment 1. (d)–(f) correspond to the moments before the occlusion, during the occlusion, and after the occlusion removal, respectively.

Once the obstruction is removed, the results of place recognition during relocalization are shown in Figure 6.7. Since p_a is located near a corner, the reference images near p_a actually have significant viewpoint changes. Additionally, there are differences in lighting between the query image and the reference images in the map, along with interference from pedestrians.

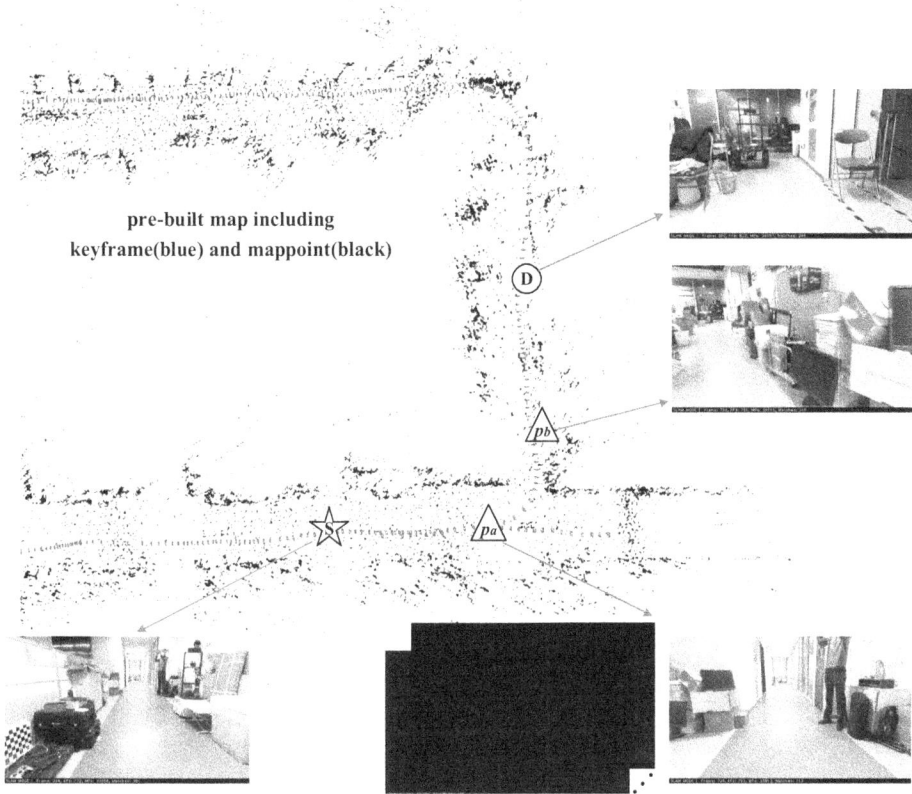

FIGURE 6.6   Offline explicit map along with grayscale images from the robot's perspective corresponding to Experiment 2. Each grayscale image displays point features and object features. At position p_a, the camera is completely occluded for approximately ten seconds.

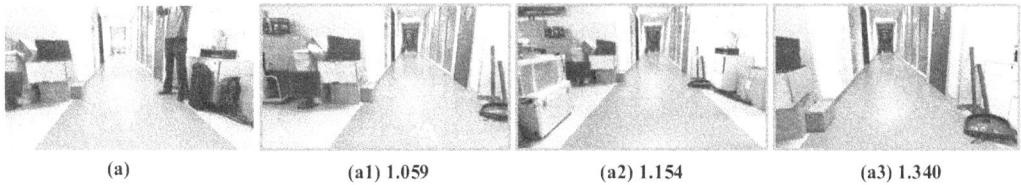

| (a) | (a1) 1.059 | (a2) 1.154 | (a3) 1.340 |

FIGURE 6.7   Query image and its top-3 retrieval results for successful relocalization at position p_a in Experiment 2. (a) Query image. (a1)–(a3) Top-3 retrieved reference images.

Despite these challenges, the correct top-3 reference images are still retrieved, as shown in Figure 6.7a1–a3. Based on these results, the robot successfully recovers its pose. Figure 6.8 provides video screenshots of the robot's navigation during Experiment 2, with Figure 6.8c–e corresponding to the moments before, during, and after the obstruction. It can be seen that the robot successfully reached its destination.

Experiment 3 examines the relocalization of the robot after it is manually dragged a certain distance during navigation. The offline explicit map and the robot's perspective images for Experiment 3 are shown in Figure 6.9. At position p_a in Figure 6.9, the camera is

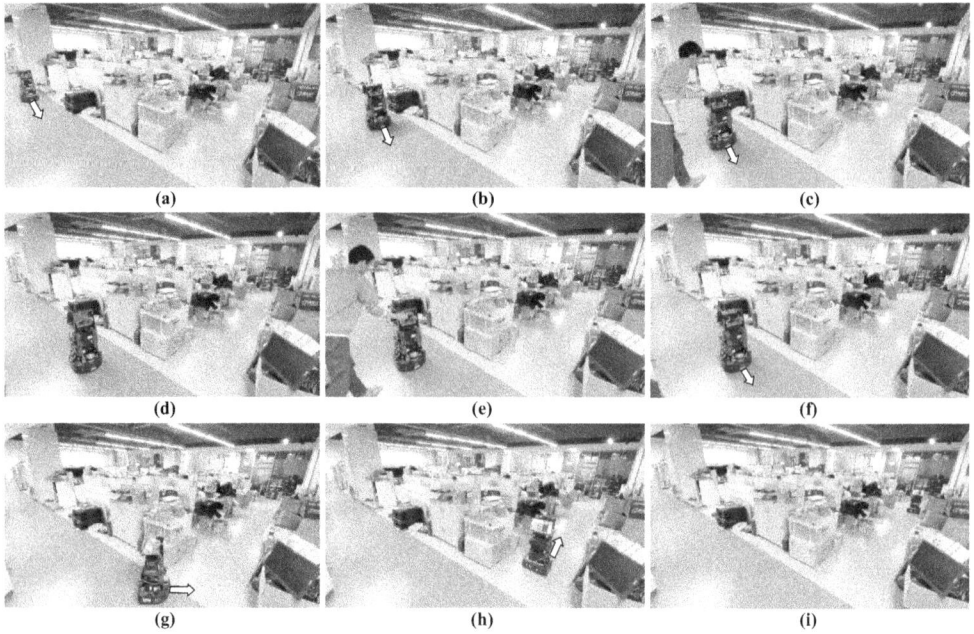

FIGURE 6.8   Video snapshots of the robot navigation in Experiment 2. (a)–(i) correspond to the snapshots at different robot positions. (c)–(e) correspond to the moments before, during, and after the occlusion, respectively.

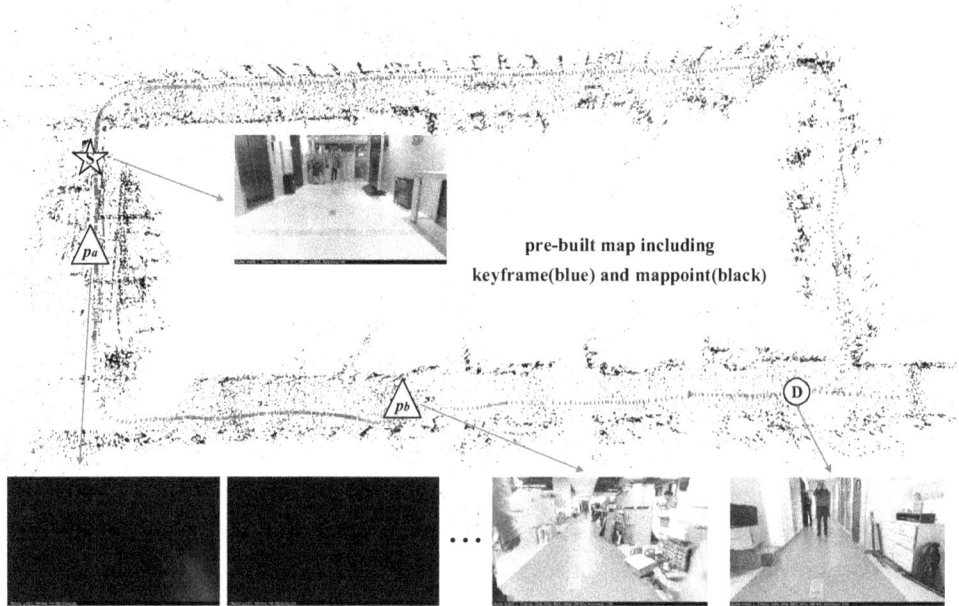

FIGURE 6.9   Offline explicit map and grayscale images from the robot's perspective in Experiment 3. The robot is manually dragged from position p_a to position p_b, with the camera completely occluded during the entire dragging process. The occlusion is removed once the robot reaches position p_b.

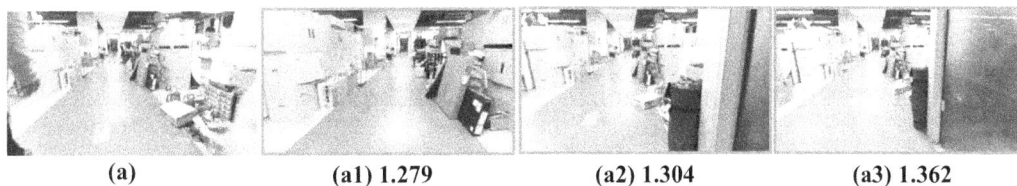

FIGURE 6.10 The query image and its top-3 retrieval results for successful localization at position p_b in Experiment 3. (a) Query image. (a1)–(a3) Top-3 retrieved reference images.

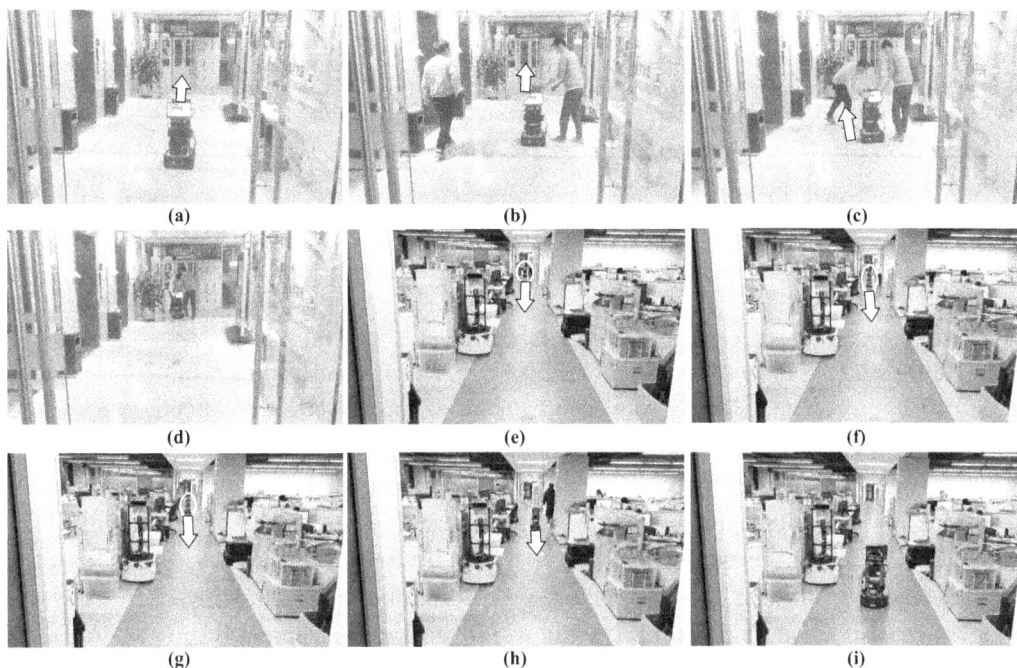

FIGURE 6.11 Video snapshots of robot navigation in Experiment 3. (a)–(i) correspond to the snapshots at different robot positions. (b) is the screenshot before the camera is obstructed. After the camera is fully obstructed, the robot is manually moved from (c) to (f).

completely obstructed, and the robot is then dragged from p_a to p_b. Upon reaching p_b, the obstruction is removed. For the query image captured by the robot at p_b (see Figure 6.10a), its top-3 retrieved images are shown in Figure 6.10a1–a3, indicating successful place recognition and subsequent acquisition of the robot's pose. Video screenshots of the robot's navigation are shown in Figure 6.11, where Figure 6.11c and f corresponds to scenes at positions p_a and p_b, respectively. After successful relocalization at p_b, the robot continues moving toward the endpoint.

6.4.2 Localization and Navigation Experiments in Corridor and Room Areas

The following considers the robot's localization and navigation in both the corridor and Room 03, where the submap of Room 03 is implicitly stored in the trained scene coordinate regression network.

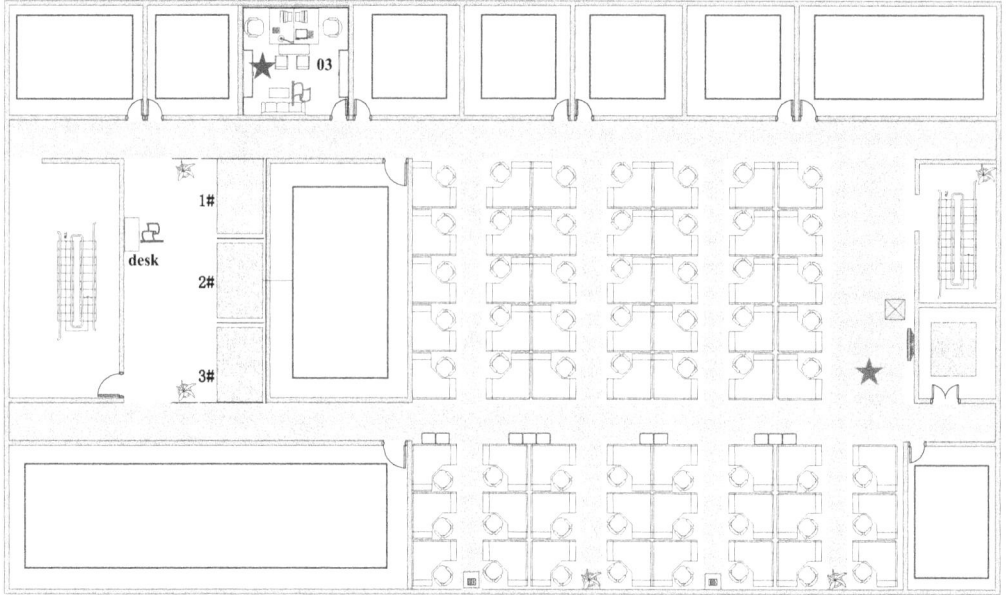

FIGURE 6.12 Experimental setups for Experiments 4 and 5. ★ in the corridor and ⌂ in Room 03 represent the starting and ending points of the robot in Experiment 4, while ★ in Room 03 and ⌂ in the corridor represent the starting and ending points of the robot in Experiment 5.

Experiment 4 simulates object delivery, as shown in Figure 6.12. The robot starts at the corridor's initial position S and is required to reach the destination D (beside the sofa in Room 03). Figure 6.13 presents the robot's perspective images and corresponding maps at four selected positions. As shown in Figure 6.13a, the robot begins moving after successful relocalization at the starting point. During its movement, the camera is fully occluded for approximately 10 seconds at position p_a in the corridor and position p_b inside the room (see Figure 6.13b and c). During these periods, the robot remains stationary. After the occlusion is removed, the robot performs relocalization. The query images with successful relocalization and corresponding results are shown in Figure 6.14. At position p_a in the corridor, the robot uses explicit submap-based relocalization. For the query image in Figure 6.14a, the position recognition results are shown in Figure 6.14a1–a3. At position p_b inside the room, the robot performs implicit submap-based relocalization. For the query image in Figure 6.14b, the scene coordinate regression network outputs the coordinate map and uncertainty map, as shown in Figure 6.14b1 and b2, respectively. The image features at successful relocalization for positions p_a and p_b are shown in Figure 6.13b and c, respectively. Finally, the robot reaches the destination D, completing the object delivery task. Video screenshots of the robot navigation process are shown in Figure 6.15. Figure 6.15c–e and h–j corresponds to the moments before, during, and after camera occlusion in the corridor and room environments, respectively. After the robot reaches the destination and stops moving, the recipient retrieves the object, as shown in Figure 6.15l.

Experiment 5 requires the robot to move from Room 03 to the elevator area and face the desk. Figure 6.16 shows the robot's perspective images and corresponding maps at four

(a)

(b)

(c)

(d)

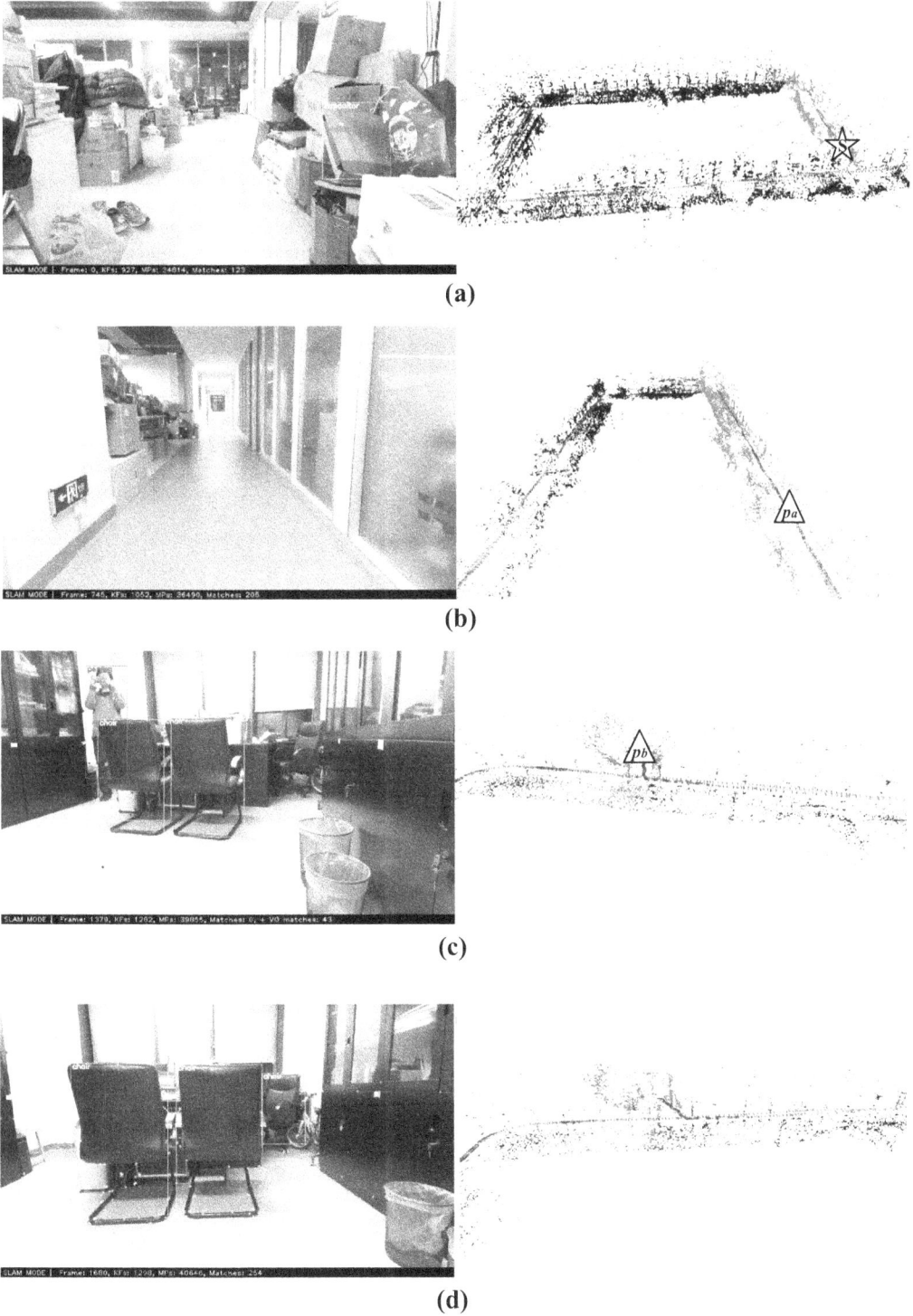

FIGURE 6.13   Images from the robot's perspective and corresponding maps at four selected locations in Experiment 4. (a) Starting position S. (b) Position p_a. (c) Position p_b. (d) Ending position D.

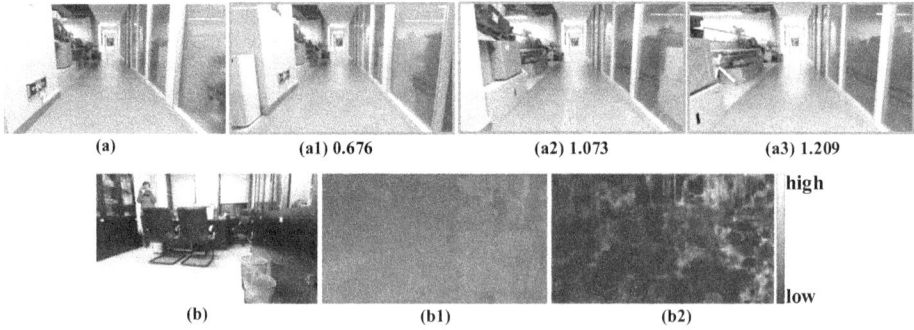

(a) (a1) 0.676 (a2) 1.073 (a3) 1.209

(b) (b1) (b2)

FIGURE 6.14 Query images with successful relocalization and corresponding results at positions p_a and p_b in Experiment 4. (a) and (a1)–(a3) show the query image at position p_a and its top-3 retrieval results; (b), (b1), and (b2) show the query image at position p_b, the 3D coordinate map and the uncertainty map visualization results output by the scene coordinate regression network, respectively.

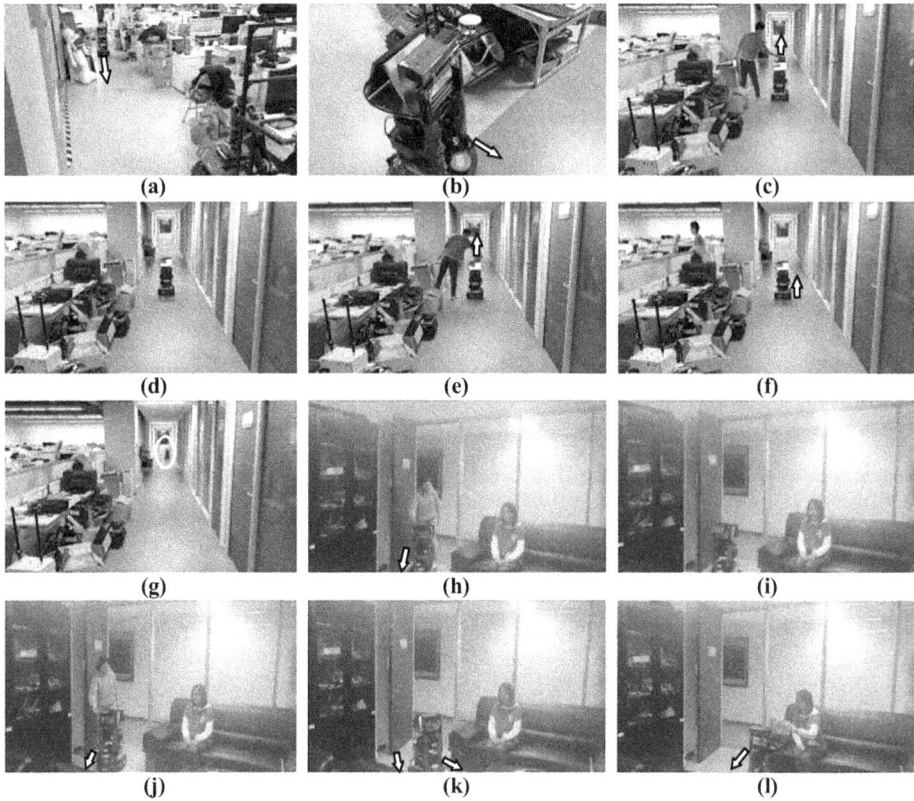

(a) (b) (c)

(d) (e) (f)

(g) (h) (i)

(j) (k) (l)

FIGURE 6.15 Video snapshots of the robot navigation in Experiment 4.

selected positions. At the starting point (see Figure 6.16a), the robot determines its initial global pose using the implicit submap-based relocalization method. The query image with successful relocalization, scene coordinate map, and uncertainty map are shown in Figure 6.17. When the robot reaches position p_a, the camera is occluded, causing it to

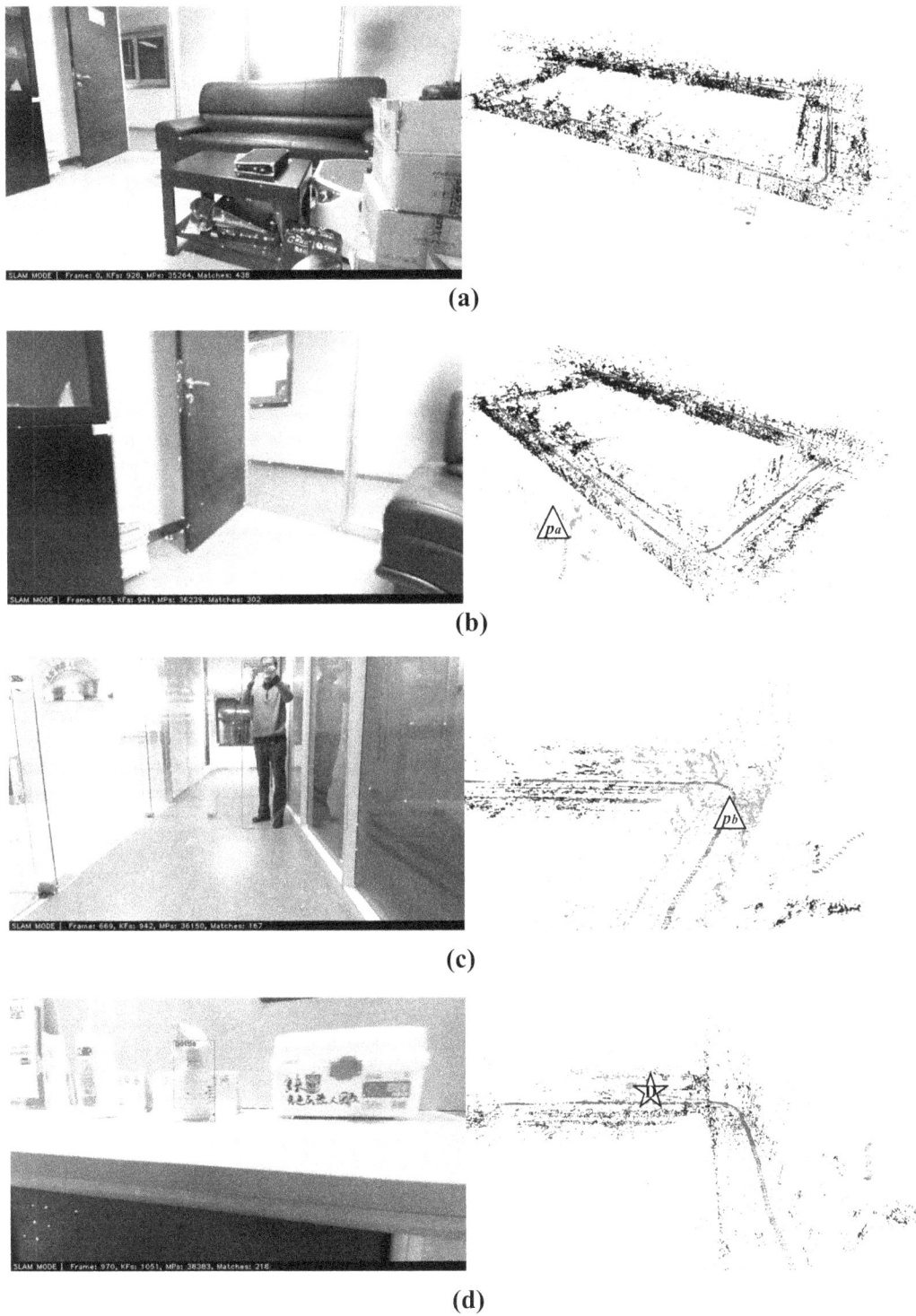

FIGURE 6.16 Images from the robot's perspective at four selected locations and the correspond-ing maps at these moments in Experiment 5. (a) Starting position S. (b) Position p_a. (c) Position p_b. (d) Destination position D.

stop moving. It is manually dragged to position p_b in the corridor, where the occlusion is removed. Figure 6.18 shows the explicit submap-based relocalization results at position p_b. The robot then resumes navigation to the destination, with video screenshots of the movement shown in Figure 6.19. The above experimental results demonstrate the effectiveness of the proposed offline hybrid map-based localization framework.

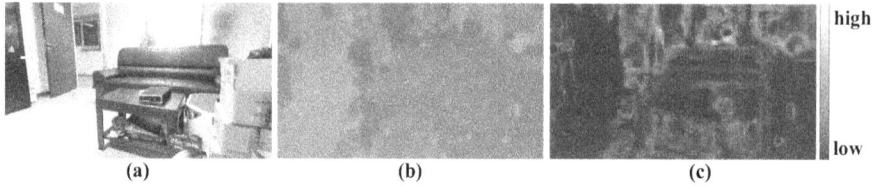

FIGURE 6.17 Query image at the starting position S from Figure 6.16, which successfully achieves relocalization, along with the 3D coordinate map and uncertainty map predicted by the scene coordinate regression network. (a) Query image at starting position S. (b) 3D coordinate map of the query image. (c) Uncertainty map of the query image.

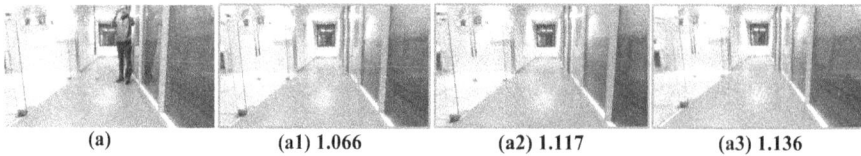

FIGURE 6.18 Query image at position p_b from Figure 6.16, which successfully achieves relocalization, along with the top-3 retrieved reference images. (a) Query image. (a1)–(a3) Top-3 retrieved reference images.

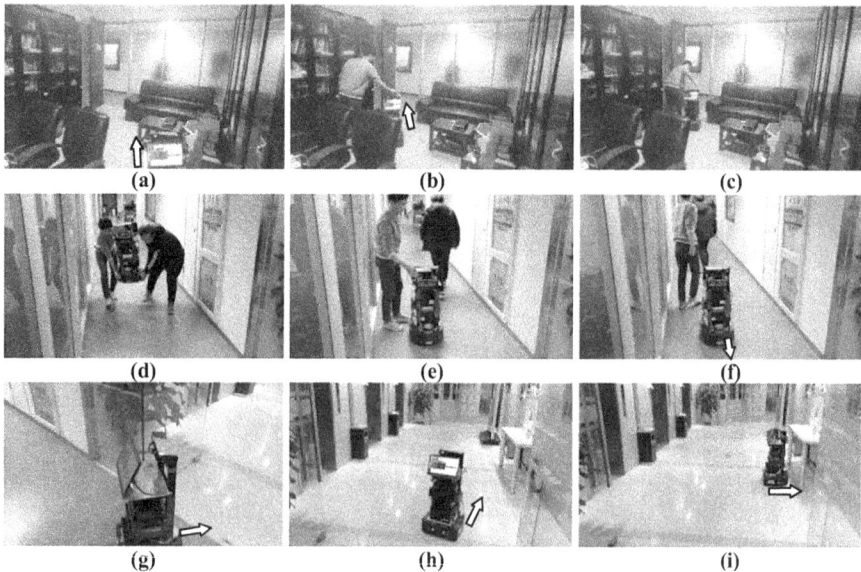

FIGURE 6.19 Video screenshots of robot navigation in Experiment 5.

6.5 CONCLUSION

In this chapter, we design an offline hybrid map-based localization architecture for service robots. Explicit submaps are constructed using the PO-SLAM and MTA methods, while implicit submaps are obtained by training the SFT-CR network. In general, tracking and localization are achieved through PO-SLAM. When tracking fails, the robot's global pose is recovered using the explicit or implicit submap-based relocalization methods. Navigation experiments in indoor corridor and room areas verify the effectiveness of the offline hybrid map-based localization framework.

REFERENCES

[1] Sattler, T., Leibe, B., & Kobbelt, L. (2011). Fast image-based localization using direct 2D-to-3D matching. In *Proceedings of the International Conference on Computer Vision*, Barcelona, Spain (pp. 667–674).

[2] Sattler, T., Leibe, B., & Kobbelt, L. (2016). Efficient & effective prioritized matching for large-scale image-based localization. *IEEE Transactions on Pattern Analysis and Machine Intelligence*, 39(9), 1744–1756.

[3] DeTone, D., Malisiewicz, T., & Rabinovich, A. (2018). Superpoint: Self-supervised interest point detection and description. In *Proceedings of the IEEE Conference on Computer Vision and Pattern Recognition Workshops*, Salt Lake City, UT, USA (pp. 224–236).

[4] Yi, K. M., Trulls, E., Lepetit, V., & Fua, P. (2016). Lift: Learned invariant feature transform. In *Proceedings of European Conference on Computer Vision*, Amsterdam, The Netherlands (pp. 467–483).

[5] Sarlin, P. E., DeTone, D., Malisiewicz, T., & Rabinovich, A. (2020). Superglue: Learning feature matching with graph neural networks. In *Proceedings of the IEEE Conference on Computer Vision and Pattern Recognition*, Seattle, WA, USA (pp. 4938–4947).

[6] Sinkhorn, R., & Knopp, P. (1967). Concerning nonnegative matrices and doubly stochastic matrices. *Pacific Journal of Mathematics*, 21(2), 343–348.

[7] Lowry, S., Sünderhauf, N., Newman, P., Leonard, J. J., Cox, D., Corke, P., & Milford, M. J. (2015). Visual place recognition: A survey. *IEEE Transactions on Robotics*, 32(1), 1–19.

[8] Zhang, X., Wang, L., & Su, Y. (2021). Visual place recognition: A survey from deep learning perspective. *Pattern Recognition*, 113, 107760.

[9] Kendall, A., Grimes, M., & Cipolla, R. (2015). PoseNet: A convolutional network for real-time 6-DoF camera relocalization. In *Proceedings of the IEEE International Conference on Computer Vision*, Santiago, Chile (pp. 2938–2946).

[10] Kendall, A., & Cipolla, R. (2016). Modelling uncertainty in deep learning for camera relocalization. In *Proceedings of IEEE International Conference on Robotics and Automation*, Stockholm, Sweden (pp. 4762–4769).

[11] Kendall, A., & Cipolla, R. (2017). Geometric loss functions for camera pose regression with deep learning. In *Proceedings of the IEEE Conference on Computer Vision and Pattern Recognition*, Honolulu, HI, USA (pp. 5974–5983).

[12] Clark, R., Wang, S., Markham, A., Trigoni, N., & Wen, H. (2017). VidLoc: A deep spatio-temporal model for 6-DoF video-clip relocalization. In *Proceedings of the IEEE Conference on Computer Vision and Pattern Recognition*, Honolulu, HI, USA (pp. 6856–6864).

[13] Wang, B., Chen, C., Lu, C. X., Zhao, P., Trigoni, N., & Markham, A. (2020, April). AtLoc: Attention guided camera localization. *Proceedings of the AAAI Conference on Artificial Intelligence*, 34(6), 10393–10401.

[14] Sattler, T., Zhou, Q., Pollefeys, M., & Leal-Taixe, L. (2019). Understanding the limitations of CNN-based absolute camera pose regression. In *Proceedings of the IEEE Conference on Computer Vision and Pattern Recognition*, Long Beach, CA, USA (pp. 3302–3312).

[15] Gao, X. S., Hou, X. R., Tang, J., & Cheng, H. F. (2003). Complete solution classification for the perspective-three-point problem. *IEEE Transactions on Pattern Analysis and Machine Intelligence*, 25(8), 930–943.

[16] Borenstein, J., & Koren, Y. (1991). The vector field histogram-fast obstacle avoidance for mobile robots. *IEEE Transactions on Robotics and Automation*, 7(3), 278–288.

CHAPTER 7

Hierarchical LiDAR Odometry via Maximum Likelihood Estimation with Tightly Associated Distributions

7.1 INTRODUCTION

Compared to the feature-based LiDAR odometry, the distribution-based solution considers more points and represents them with distribution with potential better accuracy, where the distribution for calculating likelihood function termed as sampling distribution is crucial. NDT [1,2] represents sampling distribution by target point cloud. It employs point-to-voxel correspondence, where a point in source point cloud is associated with a voxel in target point cloud. Then, each point in source point cloud is assumed to be sampled from the distribution that is described by the point set in its corresponding voxel. This voxel distribution is regarded as the sampling distribution for likelihood function calculation. A possible problem is that the sampling distribution does not accurately reflect the local geometry implied in a voxel when there are not enough points in this voxel, which will introduce noise into optimization. GICP employs point-to-point correspondence [3]. For a point in source point cloud, its nearest point in target point cloud is found to form a correspondence. By respectively searching neighbor points of these two matched points, two Gaussian distributions are obtained. The difference between two matched points is assumed to obey the difference between the two Gaussian distributions, which constitutes the sampling distribution of GICP. Although GICP utilizes both the source and target information to represent the sampling distribution, inaccurate Gaussian distribution caused by insufficient neighbor points affects the quality of the sampling distribution. Besides, point-to-point correspondence is susceptible to interference and frequent data associations during optimization are inevitable.

DOI: 10.1201/9781003643630-7

Compared to the typical distribution solutions GICP and NDT, we concern point set in source point cloud instead of a source point. For each source point set, its neighbor point set from target point cloud is searched, and we obtain a pair of the matched point sets. Actually, the matched point sets on a local region are similar in distribution, which inspires us to represent the sampling distribution of source point set using their union. This means that each source point set is only related to a sampling distribution. On one hand, source point cloud and target point cloud are tightly associated and thus the sampling distribution is expected to be more accurate, which is beneficial to better represent the local structure. On the other hand, the proposed method takes the point set as the basic unit and the resulting sparsity increases the robustness of data association. Thus, fewer data associations are involved when matching, which improves computational efficiency.

It is also worth mentioning that GICP and NDT are difficult to extend to fix-lag smoothing where there exist constraints between the observations and multi-frame pose states. Because point-to-point correspondences of GICP among multiple frames of a sliding window are tedious to find and the error term for multi-point correspondence is hard to be constructed. Together with a higher computational complexity, the expansion of GICP becomes challenging. For NDT, the form of voxel makes the data association in a sliding window easier; however, point-wise likelihood calculation causes a heavy computation burden due to massive points among the sliding window. In this chapter, points from multiple frames are organized in a form of voxel map and we treat all points from different frames falling in a voxel as a correspondence. Then, the sampling distribution followed by all points in this correspondence is described by their union. With the designed strategy that converts the likelihood terms of all points from a frame in a correspondence to a single item related to the mean and covariance of these points, the computational complexity is reduced, and an efficient fixed-lag smoothing is implemented.

The main contributions of this chapter are twofold. Firstly, a novel matching method based on maximum likelihood estimation is proposed for pose estimation with distribution-to-distribution correspondence. Specifically, the sampling distribution is represented by the union of the matched point sets in a correspondence, and both source and target point clouds are effectively exploited. As a result, local structure is reflected more accurately and the sensitivity of data association is reduced. Secondly, the proposed matching method is extended to fixed-lag smoothing with correspondence formed by the associated point sets from different frames. By decoupling the calculation of likelihood sum of all points within a point set in a correspondence from the points number, the challenge of computation complexity in existing distribution-based methods [1–3] is overcome. It can be attached to the existing LO frameworks for the improvement of performance. On this basis, a hierarchical 3D LiDAR odometry framework with low-level scan-to-map matching and high-level fixed-lag smoothing is presented and competitive results are obtained.

The rest of the chapter is organized as follows. Section 7.2 outlines the related work. In Section 7.3, the 3D LiDAR odometry is described. The experiment results are presented in Section 7.4, and Section 7.5 concludes this chapter.

7.2 REVIEW OF DISTRIBUTED-BASED METHODS

NDT [1] and GICP [3] are two typical distribution methods. On this basis, a series of improvements are put forward. Yue et al. designed a deep learning-based super-resolution network for enriching sparse 3D point cloud [4], which was combined with NDT to achieve accurate positioning in urban canyons. The challenge of localization using the sparse point cloud from low-cost LiDAR is solved. To deal with the contradiction between the voxel size and convergence behaviors, Takeuchi and Tsubouchi [5] adopted voxels of different sizes at different iteration optimization stages, improving the matching accuracy and speed of NDT in estimation. In [6], Magnusson et al. presented a point-to-multi-distribution improved NDT weighted based on trilinear interpolation for the discretization artifacts. The comparison results on a field mapping indicate that this improved NDT can increase the success rate of registration; however, it leads to longer execution time. By extending GICP with voxelization, VGICP is presented in [7]. It replaces one-to-one point distance loss with one-to-many point distance loss and uses voxels to store the means of multiple distributions. The VGICP method outperforms GICP in terms of processing speed while retaining comparable accuracy in trajectory estimation. Instead of only considering the constraint of the plane normal direction in the plane-to-plane matching like GICP, Vlaminck et al. [8] incorporated additional information to the plane directions of covariance by taking account of the curvature of local surface, which substantially reduces the odometry error. In some scenarios, such as when local geometry is a plane, covariance matrixes of GICP and NDT may degenerate. To solve this problem, Yokozuka et al. [9] improved GICP via normalizing the covariance with Frobenius norm and regularizing the cost function, which is successfully applied to LiDAR-based tracking and mapping. On this basis, symmetric KL-divergence is introduced into the cost function considering the difference in distribution shape [10], and an ultra-light LiDAR SLAM termed as LiTAMIN2 is proposed.

To further improve performance, GICP and NDT can be combined. Stoyanov et al. [11] proposed a voxel-to-voxel NDT, which voxelizes both source and target point clouds. By minimizing the sum of distances between the distributions of the paired source and target voxels, a stable registration is achieved. In addition to NDT and GICP series, there are other registration algorithms based on the Gaussian mixture module (GMM). Tabib et al. presented an on-manifold GMM matching technique enabling mapping and navigation in complex domains [12]. This method separately represents the two point clouds using two GMMs and aligns them by minimizing the squared L2 norm between two distributions. In [13], an adaptive multi-scale point matching approach is proposed by constructing a hierarchical GMM tree, which provides effective support for trajectory estimation. Note that the aforementioned distribution-based methods are not extended to the fix-lag smoothing due to the computation complexity and the problem of data association among multiple frames. In this chapter, the correspondence between two-frame distributions is extended to multi-frame case and the resulting fix-lag smoothing module can be appended to scan-to-map matching to output a more precise result.

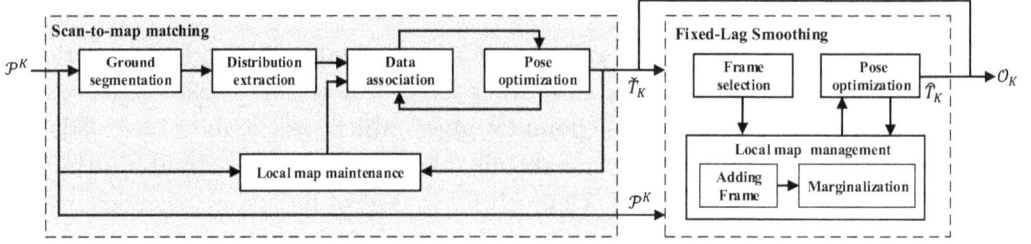

FIGURE 7.1   The pipeline of the proposed 3D LiDAR odometry HDLO.

7.3 3D LiDAR ODOMETRY

The structure of the proposed 3D LiDAR odometry termed as HDLO is shown in Figure 7.1, which mainly includes two modules: scan-to-map matching and fixed-lag smoothing. The scan-to-map matching module sequentially performs ground segmentation and distribution extraction on the input point cloud $\mathcal{P}^K$. The obtained source distributions are sent to data association to generate source-to-target distribution correspondences, based on which the maximum log-likelihood cost function is constructed for pose optimization. A preliminary pose estimate $\breve{T}_K$ is outputted. On this basis, the fixed-lag smoothing module takes $\breve{T}_K$ and $\mathcal{P}^K$ as inputs and selects some point cloud frames to construct multi-frame constraints for joint optimization. It outputs a smoother pose estimation result $\hat{T}_K$, which is fused with the estimation of scan-to-map matching to get the final localization output $\mathcal{O}_K$.

7.3.1 Scan-to-Map Matching

The scan-to-map matching module receives 3D point cloud directly from the LiDAR sensor and provides the pose estimation result of the newest frame as the initial value for the fixed-lag smoothing module.

7.3.1.1 Ground Segmentation

In the distribution-based methods, the Gaussian distribution is commonly used. Our method employs point sets to describe Gaussian distributions; however, the extracted point sets nearby the ground maybe contain both ground points and non-ground points, leading to a vertical structure which is hard to be represented with a Gaussian distribution model. To avoid this problem, a ground segmentation preprocessing is first executed for better distribution representation.

We denote the downsampled point cloud of current frame K represented in LiDAR coordinate system as $\mathcal{P}^K$. The ground segmentation algorithm in reference [14] is used to divide $\mathcal{P}^K$ into ground points $\mathcal{P}_g^K$ and non-ground points $\mathcal{P}_{ng}^K$. The core of this algorithm is to segment the point cloud into different patches according to x and y coordinates and then fit the corresponding planes. After that, ground points and non-ground points are separated. The subsequent operations in the scan-to-map matching module shall be, respectively, performed on $\mathcal{P}_g^K$ and $\mathcal{P}_{ng}^K$.

7.3.1.2 Distribution Extraction and Data Association

$\mathcal{P}_g^K$ and $\mathcal{P}_{ng}^K$ are used as source point clouds that are aligned to their respective target point clouds. The target point clouds refer to local maps, which are composed of downsampled ground points and non-ground points from past selected l frames. Each local map is represented with voxel grids in the global frame, which coincides with the first LiDAR frame. As mentioned above, we need to search for Gaussian distribution correspondences between each source point cloud and its target point cloud, where the distribution is described by a set of points with its mean and covariance. For the simplicity of description, we will not distinguish point set from distribution.

To extract Gaussian distributions, we firstly voxelize the source point cloud and record all voxel centroids, which serve as candidate locations of distributions, where a voxel centroid refers to the mean of all points in a voxel. The voxel size for ground point cloud can be set larger than that of non-ground point cloud since the LiDAR ring is sparser in ground. Then, a KD-tree of a source point cloud is built to find the point set of neighbors for each centroid by radius search. Considering the change of point cloud density from near to far, a search strategy with varying radius is adopted. The initial search radius is set to half of the voxel size. When the number of points in the searched point set is inadequate, the search radius shall be extended to 1.5 times. For each searched point set with enough points, its mean and covariance are calculated, obtaining a Gaussian distribution. Besides, considering some distant local regions but with good geometry, such as tree, pole, etc., the searched point set with less quantity is also concerned. In this case, the geometry σ_g is calculated by $\dfrac{\lambda_1 + \lambda_2}{\lambda_1 + \lambda_2 + \lambda_3}$ $(\lambda_1 \geq \lambda_2 \geq \lambda_3)$, where $\lambda_i (1 \leq i \leq 3)$ are the eigenvalues of the covariance. The point set with a larger geometry shall be kept to form a distribution.

For each ground point set, its linearity $\sigma_l = \dfrac{\lambda_1 - \lambda_2}{\lambda_1}$ [15] is calculated, and those that distribute linearly are directly excluded to better keep the planar characteristics of ground point sets. To avoid the overlapping of distributions, those voxel centroids that fall close to the centers of the extracted distributions will no longer be considered. Algorithm 7.1 describes the detailed distribution extraction process, where I_g and I_{ng} are the sets of distributions corresponding to ground and non-ground point clouds, respectively. r_g and r_{ng} are the initial search radiuses. RadiusSearch $(c_i, r, \mathbb{T}_P)$ is used to search a set of points within the radius r at the center of c_i based on the KD-tree $\mathbb{T}_P$. Taking a frame in the KITTI sequence 00 as an example, Figure 7.2 illustrates the results of distributions extracted from ground points and non-ground points.

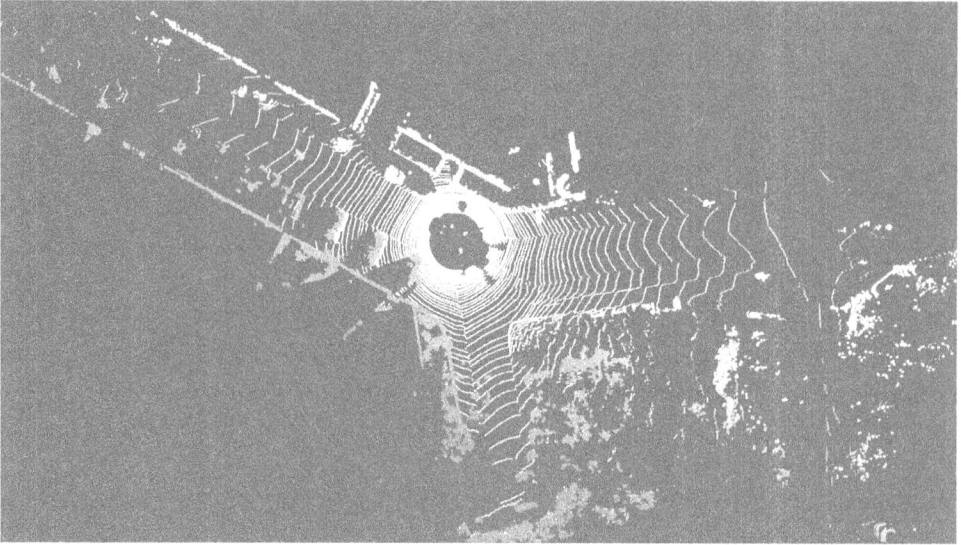

(a)

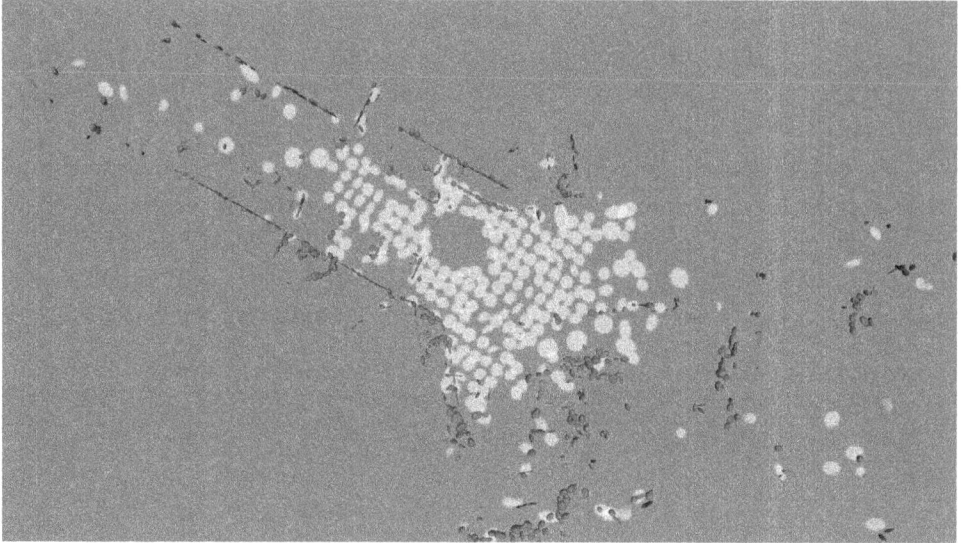

(b)

FIGURE 7.2   An illustration of Gaussian distribution extraction. (a) A frame of point cloud in the KITTI sequence 00. (b) Gaussian distributions extracted from ground points and non-ground points.

Algorithm 7.1 Distribution Extraction

Input: ground point cloud $\mathcal{P}_g^K$ and non-ground point cloud $\mathcal{P}_{ng}^K$

Output: distribution sets I_g and I_{ng}

1   $I_g = \varnothing, I_{ng} = \varnothing$;
2   $r_g = 0.5, r_{ng} = 1.0$;

3　　**for** $(\mathcal{P},r,I)\in\left\{\left(\mathcal{P}_g^K,r_g,I_g\right),\left(\mathcal{P}_{ng}^K,r_{ng},I_{ng}\right)\right\}$ **do**
4　　　Voxelize $\mathcal{P}$ and extract candidate centroids C;
5　　　Build a KD-tree $\mathbb{T}_{\mathcal{P}}$ for $\mathcal{P}$ and a KD-tree $\mathbb{T}_C$ for C;
6　　　**for** c_i in C **do**
7　　　　**if** c_i is deactivated **do**
8　　　　　**continue**;
9　　　　**end if**
10　　　　$S = \text{RadiusSearch}\left(c_i,r,\mathbb{T}_{\mathcal{P}}\right)$;
11　　　　**if** $|S| < N_p$ **do**
12　　　　　$S = \text{RadiusSearch}\left(c_i,1.5r,\mathbb{T}_{\mathcal{P}}\right)$;
13　　　　**end if**
14　　　　Calculate mean and covariance of S;
15　　　　Calculate σ_g and σ_l;
16　　　　**if** $|S| > N_p$ or ($|S| > N_c$ and $\sigma_g > t_g$) **do**
17　　　　　**if** $\mathcal{P} == \mathcal{P}_g^K$ and $\sigma_l > t_l$ **do**
18　　　　　　**continue**;
19　　　　　**end if**
20　　　　　$I = I \cup S$;
21　　　　　$Q = \text{RadiusSearch}\left(c_i, 1.5\max_{s_a \in S}\left(\|s_a - c_i\|_2\right), \mathbb{T}_C\right)$;
22　　　　　deactivate each centroid in Q;
23　　　　**end if**
24　　　**end for**
25　　**end for**
26　**return**

The purpose of data association is to search for a distribution $\mathcal{T}_i$ in the target point cloud in correspondence with each distribution $\mathcal{S}_i$ extracted from source point cloud. The centroid of $\mathcal{S}_i$ is firstly projected to the global frame with the transformation T_K from the Kth LiDAR frame to the global frame, where $T_K = (R_K, t_K) \in SO(3) \times \mathbb{R}^3$. Then, all points adjacent to the projection position in the local map constitute the distribution $\mathcal{T}_i$. Note that the points in a larger neighborhood of the projection position are searched in the initialization of scan-to-map matching to deal with the instability of initial data association. Examples of data association of two Gaussian distributions on local regions of the KITTI sequence 00 are provided in Figure 7.3, where (a) and (b) reflect the local regions of a tree trunk and a wall, respectively. One can see that the extracted distribution in target point cloud is similar to that in source point cloud.

7.3.1.3 Pose Optimization

With data association as described above, two sets of matched distributions $\mathcal{S} = \left\{ \mathcal{S}_i \mid \mathcal{S}_i \sim \mathcal{N}\left(\mu_i^{\text{S}}, \Sigma_i^{\text{S}}\right), 1 \leq i \leq N \right\}$ and $\mathcal{T} = \left\{ \mathcal{T}_i \mid \mathcal{T}_i \sim \mathcal{N}\left(\mu_i^{\text{T}}, \Sigma_i^{\text{T}}\right), 1 \leq i \leq N \right\}$ are obtained, where $\mathcal{S}_i$ and $\mathcal{T}_i$ are the ith matched distributions, and N is the number of

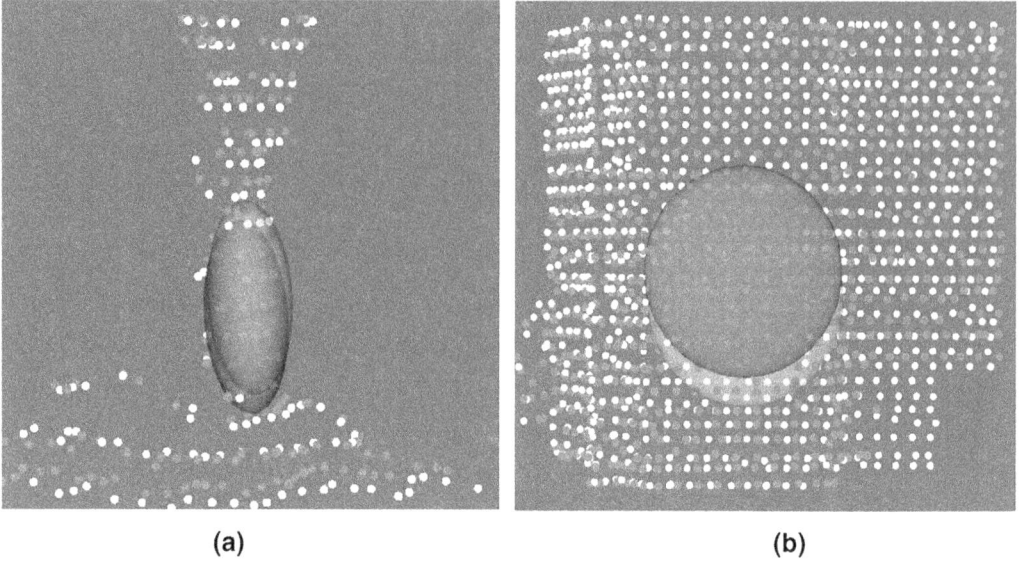

(a)   **(b)**

FIGURE 7.3 Examples of data association of two Gaussian distributions on local regions of the KITTI sequence 00. (a) Tree trunk. (b) Wall. The ellipsoids are the extracted Gaussian distributions from the corresponding point clouds.

correspondences. $\mathcal{N}(\mu, \Sigma)$ represents Gaussian distribution in which μ and Σ refer to its mean and covariance, respectively. For each $\mathcal{S}_i$, it is transformed to the global frame to get $\mathcal{Q}_i \sim \mathcal{N}(\mu_i^Q, \Sigma_i^Q)$, where $\mu_i^Q = R_K \mu_i^S + t_K$ and $\Sigma_i^Q = R_K \Sigma_i^S R_K^T$. Because $\mathcal{S}_i$ and $\mathcal{T}_i$ are matched, it can be reasonably considered that $\mathcal{Q}_i$ and $\mathcal{T}_i$ have similar distributions, that is, μ_i^Q is close to μ_i^T in Euclidean distance and the scales of the eigenvalues and the directions of the eigenvectors of Σ_i^Q and Σ_i^T are similar. We consider they are sampled from the same distribution $\mathcal{C}_i \sim \mathcal{N}(\mu_i^C, \Sigma_i^C)$ described by the union of $\mathcal{Q}_i$ and $\mathcal{T}_i$. Therefore, the odometry estimation can be modeled as an optimization problem shown in (7.1) by maximizing the sum of log-likelihood function of $\mathcal{C}_i$ $(1 \le i \le N)$:

$$\max_{T_K} -\frac{1}{2} \sum_{i=1}^{N} \frac{w_i}{M_i^Q} \sum_{j=1}^{M_i^Q} e_{ij}^T (T_K) \left(\Sigma_i^C \right)^{-1} e_{ij} (T_K) \tag{7.1}$$

where

$$e_{ij}(T_K) = R_K \cdot p_{ij}^S + t_K - \mu_i^C \tag{7.2}$$

$$\mu_i^C = \frac{M_i^Q \mu_i^Q + M_i^T \mu_i^T}{M_i^Q + M_i^T} \tag{7.3}$$

$$\Sigma_i^C = \frac{M_i^Q \left(\Sigma_i^Q + \mu_i^Q \left(\mu_i^Q \right)^T \right) + M_i^T \left(\Sigma_i^T + \mu_i^T \left(\mu_i^T \right)^T \right)}{M_i^Q + M_i^T} - \mu_i^C \left(\mu_i^C \right)^T \tag{7.4}$$

M_i^Q and M_i^T are the numbers of points in Q_i and T_i, and p_{ij}^S refers to the jth point in S_i. $w_i = \sqrt{M_i^Q - 5} \cdot d_{Q_i T_i} d_{Q_i C_i} d_{T_i C_i}$ is the weight to measure the importance of the ith log-likelihood function, where $d_{Q_i T_i}$, $d_{Q_i C_i}$, and $d_{T_i C_i}$ are used to evaluate the similarity between two of the distributions Q_i, T_i, and C_i. Following the correlation matrix distance of two matrixes in reference [16], a symmetrical form is presented to represent the similarity of two distributions. $d_{Q_i T_i}$ is given by

$$d_{Q_i T_i} = \frac{1}{2} * \left(2 - \frac{\mathrm{Tr}\left(\left(\Sigma_i^Q\right)^{-1} \Sigma_i^T\right)}{\left\|\left(\Sigma_i^Q\right)^{-1}\right\| \cdot \left\|\Sigma_i^T\right\|} - \frac{\mathrm{Tr}\left(\left(\Sigma_i^T\right)^{-1} \Sigma_i^Q\right)}{\left\|\left(\Sigma_i^T\right)^{-1}\right\| \cdot \left\|\Sigma_i^Q\right\|} \right) \tag{7.5}$$

where $\|\cdot\|$ and $\mathrm{Tr}(\cdot)$ describe Frobenius norm and trace, respectively. Similarly, $d_{Q_i C_i}$ and $d_{T_i C_i}$ are acquired.

The above problem is a complicated non-convex optimization one, which can be solved by numerical optimization in an iteration way. It is noted that μ_i^C and Σ_i^C are also the functions of T_K, which implies that (7.1) is not a basic least square problem. In this chapter, a two-step policy is adopted. In each iteration, μ_i^C and Σ_i^C are first calculated with the latest T_K, and then (7.1) turns into a common problem that can be directly solved by Gauss-Newton algorithm. The rotation matrix R_K is optimized in Lie algebra space $\mathfrak{so}(3)$ instead of manifold $SO(3)$ with the right perturbation model [17]. Thus, the derivative of $e_{ij}(T_K)$ with respect to the pose T_K is as follows:

$$J_{ij} = \frac{\partial e_{ij}(T_K)}{\partial T_K} = \left[\begin{array}{cc} -R_K \left[p_{ij}^S \right]^{\wedge} & I_{3\times3} \end{array} \right] \tag{7.6}$$

where $[\cdot]^{\wedge}$ is skew symmetric matrix. $J_{ij}(*, 1:3)$ and $J_{ij}(*, 4:6)$ describe the derivatives with respect to the right perturbation in the tangent plane of R_K, and the translation t_K. On this basis, the incremental equation for optimization can be obtained by

$$\sum_{i=1}^{N} \frac{w_i}{M_i^Q} \sum_{j=1}^{M_i^Q} J_{ij}^T \left(\Sigma_i^C\right)^{-1} J_{ij} \delta x = -\sum_{i=1}^{N} \frac{w_i}{M_i^Q} \sum_{j=1}^{M_i^Q} J_{ij}^T \left(\Sigma_i^C\right)^{-1} e_{ij} \tag{7.7}$$

where $\delta x \in \mathfrak{so}(3) \times \mathbb{R}^3$ represents the increment of T_K. Clearly, the complexity of (7.7) is related to not only the number N of the matched distributions but also the number of points within each distribution, resulting in a heavy computational burden. To solve this problem, a decoupling strategy is proposed to eliminate the dependency on the point number, which means that $\sum_{j=1}^{M_i^Q} J_{ij}^T \left(\Sigma_i^C\right)^{-1} e_{ij}$ and $\sum_{j=1}^{M_i^Q} J_{ij}^T \left(\Sigma_i^C\right)^{-1} J_{ij}$ in the inner layer of (7.7) are turned into non-sum forms independent of j. The decoupling results are given in formulas (7.8) and (7.9):

$$\sum_{j=1}^{M_i^Q} J_{ij}^T \left(\Sigma_i^C\right)^{-1} e_{ij} = \begin{bmatrix} \Omega_1 + M_i^Q \left[\mu_i^S\right]^\wedge \cdot R_K^T \left(\Sigma_i^C\right)^{-1} \left(t_K - \mu_i^C\right) \\ M_i^Q \left(\Sigma_i^C\right)^{-1} \left(R_K \mu_i^S + t_K - \mu_i^C\right) \end{bmatrix} \tag{7.8}$$

$$\sum_{j=1}^{M_i^Q} J_{ij}^T \left(\Sigma_i^C\right)^{-1} J_{ij} = \begin{bmatrix} \Omega_2 & M_i^Q \left[\mu_i^S\right]^\wedge \cdot R_K^T \left(\Sigma_i^C\right)^{-1} \\ -M_i^Q \left(\Sigma_i^C\right)^{-1} R_K \left[\mu_i^S\right]^\wedge & M_i^Q \left(\Sigma_i^C\right)^{-1} \end{bmatrix} \tag{7.9}$$

where $\Omega_1 = \begin{bmatrix} V_1 \odot W \\ V_2 \odot W \\ V_3 \odot W \end{bmatrix}$, $\Omega_2 = \begin{bmatrix} V_4 \odot W & V_5 \odot W & V_6 \odot W \\ V_5 \odot W & V_7 \odot W & V_8 \odot W \\ V_6 \odot W & V_8 \odot W & V_9 \odot W \end{bmatrix}$, and $W = R_K^T \left(\Sigma_i^C\right)^{-1} R_K$.

The symbol $\odot$ is the element-wise production sum, following the rule $V \odot W = \sum_{i_r j_c} v_{i_r j_c} w_{i_r j_c}$, where $v_{i_r j_c}$ and $w_{i_r j_c}$ are the elements in the i_r^{th} row and j_c^{th} column of the matrixes V and W, respectively. $V_1 \sim V_9$ are the matrixes composed of the elements in $\Gamma = \sum_{j=1}^{M_i^Q} p_{ij}^S \left(p_{ij}^S\right)^T$.

The derivation of formulas (7.8) is as follows:

$$\sum_{j=1}^{M_i^Q} J_{ij}^T \left(\Sigma_i^C\right)^{-1} e_{ij} = \sum_{j=1}^{M_i^Q} \begin{bmatrix} \left[p_{ij}^S\right]^\wedge R_K^T \\ I_{3\times3} \end{bmatrix} \cdot \left(\Sigma_i^C\right)^{-1} \cdot \left(R_K p_{ij}^S + t_K - \mu_i^C\right)$$

$$= \begin{bmatrix} \sum_{j=1}^{M_i^Q} \left[p_{ij}^S\right]^\wedge R_K^T \left(\Sigma_i^C\right)^{-1} R_K p_{ij}^S + \sum_{j=1}^{M_i^Q} \left[p_{ij}^S\right]^\wedge \cdot R_K^T \left(\Sigma_i^C\right)^{-1} \left(t_K - \mu_i^C\right) \\ \left(\Sigma_i^C\right)^{-1} R_K \sum_{j=1}^{M_i^Q} p_{ij}^S + \left(\Sigma_i^C\right)^{-1} \sum_{j=1}^{M_i^Q} \left(t_K - \mu_i^C\right) \end{bmatrix} \tag{7.10}$$

$$= \begin{bmatrix} \sum_{j=1}^{M_i^Q} \left[p_{ij}^S\right]^\wedge R_K^T \left(\Sigma_i^C\right)^{-1} R_K p_{ij}^S + M_i^Q \left[\mu_i^S\right]^\wedge \cdot R_K^T \left(\Sigma_i^C\right)^{-1} \left(t_K - \mu_i^C\right) \\ M_i^Q \left(\Sigma_i^C\right)^{-1} \left(R_K \mu_i^S + t_K - \mu_i^C\right) \end{bmatrix}$$

$u_{1\times3}U_{3\times3}v_{3\times1}$ can be written as $u_{1\times3}^{T}v_{3\times1}^{T} \odot U_{3\times3}$. Then, the item $\displaystyle\sum_{j=1}^{M_i^Q}\left[p_{ij}^{S}\right]^{\wedge}R_K^{T}\left(\Sigma_i^{C}\right)^{-1}R_K p_{ij}^{S}$ in

(7.10) becomes

$$\sum_{j=1}^{M_i^Q}\left[p_{ij}^{S}\right]^{\wedge}R_K^{T}\left(\Sigma_i^{C}\right)^{-1}R_K p_{ij}^{S}$$

$$=\begin{bmatrix}\left(\displaystyle\sum_{j=1}^{M_i^Q}\left(\left[p_{ij}^{S}\right]^{\wedge}\right)_{1\star}^{T}\cdot\left(p_{ij}^{S}\right)^{T}\right)\odot\left(R_K^{T}\left(\Sigma_i^{C}\right)^{-1}R_K\right)\\[2em]\left(\displaystyle\sum_{j=1}^{M_i^Q}\left(\left[p_{ij}^{S}\right]^{\wedge}\right)_{2\star}^{T}\cdot\left(p_{ij}^{S}\right)^{T}\right)\odot\left(R_K^{T}\left(\Sigma_i^{C}\right)^{-1}R_K\right)\\[2em]\left(\displaystyle\sum_{j=1}^{M_i^Q}\left(\left[p_{ij}^{S}\right]^{\wedge}\right)_{3\star}^{T}\cdot\left(p_{ij}^{S}\right)^{T}\right)\odot\left(R_K^{T}\left(\Sigma_i^{C}\right)^{-1}R_K\right)\end{bmatrix}\qquad(7.11)$$

$$=\begin{bmatrix}\left(\displaystyle\sum_{j=1}^{M_i^Q}\left(\left[p_{ij}^{S}\right]^{\wedge}\right)_{1\star}^{T}\cdot\left(p_{ij}^{S}\right)^{T}\right)\odot W\\[2em]\left(\displaystyle\sum_{j=1}^{M_i^Q}\left(\left[p_{ij}^{S}\right]^{\wedge}\right)_{2\star}^{T}\cdot\left(p_{ij}^{S}\right)^{T}\right)\odot W\\[2em]\left(\displaystyle\sum_{j=1}^{M_i^Q}\left(\left[p_{ij}^{S}\right]^{\wedge}\right)_{3\star}^{T}\cdot\left(p_{ij}^{S}\right)^{T}\right)\odot W\end{bmatrix}$$

where $\left(\left[p_{ij}^{S}\right]^{\wedge}\right)_{a\star}^{T}$ represents the transpose of the ath row of $\left[p_{ij}^{S}\right]^{\wedge}$. Denote p_{ij}^{S} as

$\begin{bmatrix}x_{ij}&y_{ij}&z_{ij}\end{bmatrix}^{T}$ and take $\displaystyle\sum_{j=1}^{M_i^Q}\left(\left[p_{ij}^{S}\right]^{\wedge}\right)_{1\star}^{T}\cdot\left(p_{ij}^{S}\right)^{T}$ in the first row of (7.11) as an example:

$$\sum_{j=1}^{M_i^Q} \left(\left[p_{ij}^S \right]^{\wedge} \right)_{1^*}^T \cdot \left(p_{ij}^S \right)^T = \sum_{j=1}^{M_i^Q} \begin{bmatrix} 0 \\ -z_{ij} \\ y_{ij} \end{bmatrix} \begin{bmatrix} x_{ij} & y_{ij} & z_{ij} \end{bmatrix}$$

$$= \sum_{j=1}^{M_i^Q} \begin{bmatrix} 0 & 0 & 0 \\ -x_{ij}z_{ij} & -y_{ij}z_{ij} & -z_{ij}z_{ij} \\ x_{ij}y_{ij} & y_{ij}y_{ij} & y_{ij}z_{ij} \end{bmatrix}$$

$$= \begin{bmatrix} 0 & 0 & 0 \\ -xz_i & -yz_i & -zz_i \\ xy_i & yy_i & yz_i \end{bmatrix}$$

$$= \begin{bmatrix} 0 & 0 & 0 \\ -\Gamma_{13} & -\Gamma_{23} & -\Gamma_{33} \\ \Gamma_{12} & \Gamma_{22} & \Gamma_{23} \end{bmatrix} = V_1$$

(7.12)

where

$$\Gamma = \sum_{j=1}^{M_i^Q} p_{ij}^S \left(p_{ij}^S \right)^T = \sum_{j=1}^{M_i^Q} \begin{bmatrix} x_{ij} \\ y_{ij} \\ z_{ij} \end{bmatrix} \begin{bmatrix} x_{ij} & y_{ij} & z_{ij} \end{bmatrix}$$

$$= \sum_{j=1}^{M_i^Q} \begin{bmatrix} x_{ij}x_{ij} & x_{ij}y_{ij} & x_{ij}z_{ij} \\ x_{ij}y_{ij} & y_{ij}y_{ij} & y_{ij}z_{ij} \\ x_{ij}z_{ij} & y_{ij}z_{ij} & z_{ij}z_{ij} \end{bmatrix}$$

(7.13)

$$= \begin{bmatrix} xx_i & xy_i & xz_i \\ xy_i & yy_i & yz_i \\ xz_i & yz_i & zz_i \end{bmatrix}$$

$$= \begin{bmatrix} \Gamma_{11} & \Gamma_{12} & \Gamma_{13} \\ \Gamma_{12} & \Gamma_{22} & \Gamma_{23} \\ \Gamma_{13} & \Gamma_{23} & \Gamma_{33} \end{bmatrix}$$

Similarly, we have

$$\sum_{j=1}^{M_i^Q} \left(\left[p_{ij}^S \right]^{\wedge} \right)_{2^*}^T \cdot \left(p_{ij}^S \right)^T = \begin{bmatrix} \Gamma_{13} & \Gamma_{23} & \Gamma_{33} \\ 0 & 0 & 0 \\ -\Gamma_{11} & -\Gamma_{12} & -\Gamma_{13} \end{bmatrix} = V_2$$

(7.14)

$$\sum_{j=1}^{M_i^Q} \left(\left[p_{ij}^S\right]^{\wedge}\right)_{3^*}^T \cdot \left(p_{ij}^S\right)^T = \begin{bmatrix} -\Gamma_{12} & -\Gamma_{22} & -\Gamma_{23} \\ \Gamma_{11} & \Gamma_{12} & \Gamma_{13} \\ 0 & 0 & 0 \end{bmatrix} = V_3 \tag{7.15}$$

Therefore, the formula (7.11) is transformed as (7.16), which is further inserted into (7.10) to get (7.8):

$$\sum_{j=1}^{M_i^Q} \left[p_{ij}^S\right]^{\wedge} R_K^T \left(\Sigma_i^C\right)^{-1} R_K p_{ij}^S = \begin{bmatrix} V_1 \odot W \\ V_2 \odot W \\ V_3 \odot W \end{bmatrix} = \Omega_1 \tag{7.16}$$

The proof of formula (7.9) is as follows:

$$\sum_{j=1}^{M_i^Q} J_{ij}^T \left(\Sigma_i^C\right)^{-1} J_{ij} = \sum_{j=1}^{M_i^Q} \begin{bmatrix} \left[p_{ij}^S\right]^{\wedge} R_K^T \\ I_{3\times3} \end{bmatrix} \cdot \left(\Sigma_i^C\right)^{-1} \cdot \begin{bmatrix} -R_K\left[p_{ij}^S\right]^{\wedge} & I_{3\times3} \end{bmatrix}$$

$$= \sum_{j=1}^{M_i^Q} \begin{bmatrix} -\left[p_{ij}^S\right]^{\wedge} R_K^T \left(\Sigma_i^C\right)^{-1} R_K\left[p_{ij}^S\right]^{\wedge} & \left[p_{ij}^S\right]^{\wedge} R_K^T \left(\Sigma_i^C\right)^{-1} \\ -\left(\Sigma_i^C\right)^{-1} R_K\left[p_{ij}^S\right]^{\wedge} & \left(\Sigma_i^C\right)^{-1} \end{bmatrix} \tag{7.17}$$

$$= \begin{bmatrix} -\sum_{j=1}^{M_i^Q}\left[p_{ij}^S\right]^{\wedge} R_K^T \left(\Sigma_i^C\right)^{-1} R_K\left[p_{ij}^S\right]^{\wedge} & M_i^Q\left[\mu_i^S\right]^{\wedge} R_K^T \left(\Sigma_i^C\right)^{-1} \\ -M_i^Q\left(\Sigma_i^C\right)^{-1} R_K\left[\mu_i^S\right]^{\wedge} & M_i^Q\left(\Sigma_i^C\right)^{-1} \end{bmatrix}$$

The element in the pth row and q^{th} column of $-\sum_{j=1}^{M_i^Q}\left[p_{ij}^S\right]^{\wedge} R_K^T \left(\Sigma_i^C\right)^{-1} R_K\left[p_{ij}^S\right]^{\wedge}$ can be rewritten as $\left(\sum_{j=1}^{M_i^Q}\left(\left[p_{ij}^S\right]^{\wedge}\right)_{p^*}^T \left(\left[p_{ij}^S\right]^{\wedge}\right)_{q^*}\right) \odot \left(R_K^T\left(\Sigma_i^C\right)^{-1} R_K\right) = \left(\sum_{j=1}^{M_i^Q}\left(\left[p_{ij}^S\right]^{\wedge}\right)_{p^*}^T \left(\left[p_{ij}^S\right]^{\wedge}\right)_{q^*}\right) \odot W.$

Considering the symmetry of the whole matrix, we only need to calculate the elements in its upper triangle part. Similar to the calculation in (7.12), $\sum_{j=1}^{M_i^Q}\left(\left[p_{ij}^S\right]^{\wedge}\right)_{p^*}^T \left(\left[p_{ij}^S\right]^{\wedge}\right)_{q^*}$ can be acquired as follows:

$$\sum_{j=1}^{M_i^Q} \left(\left[p_{ij}^S\right]^{\wedge}\right)_{1^*}^T \left(\left[p_{ij}^S\right]^{\wedge}\right)_{1^*} = \begin{bmatrix} 0 & 0 & 0 \\ 0 & \Gamma_{33} & -\Gamma_{23} \\ 0 & -\Gamma_{23} & \Gamma_{22} \end{bmatrix} = V_4 \tag{7.18}$$

$$\sum_{j=1}^{M_i^Q}\left(\left[p_{ij}^S\right]^\wedge\right)_{1^*}^T\left(\left[p_{ij}^S\right]^\wedge\right)_{2^*}=\begin{bmatrix}0&0&0\\-\Gamma_{33}&0&\Gamma_{13}\\\Gamma_{23}&0&-\Gamma_{12}\end{bmatrix}=V_5 \tag{7.19}$$

$$\sum_{j=1}^{M_i^Q}\left(\left[p_{ij}^S\right]^\wedge\right)_{1^*}^T\left(\left[p_{ij}^S\right]^\wedge\right)_{3^*}=\begin{bmatrix}0&0&0\\\Gamma_{23}&-\Gamma_{13}&0\\-\Gamma_{22}&\Gamma_{12}&0\end{bmatrix}=V_6 \tag{7.20}$$

$$\sum_{j=1}^{M_i^Q}\left(\left[p_{ij}^S\right]^\wedge\right)_{2^*}^T\left(\left[p_{ij}^S\right]^\wedge\right)_{2^*}=\begin{bmatrix}\Gamma_{33}&0&-\Gamma_{13}\\0&0&0\\-\Gamma_{13}&0&\Gamma_{11}\end{bmatrix}=V_7 \tag{7.21}$$

$$\sum_{j=1}^{M_i^Q}\left(\left[p_{ij}^S\right]^\wedge\right)_{2^*}^T\left(\left[p_{ij}^S\right]^\wedge\right)_{3^*}=\begin{bmatrix}-\Gamma_{23}&\Gamma_{13}&0\\0&0&0\\\Gamma_{12}&-\Gamma_{11}&0\end{bmatrix}=V_8 \tag{7.22}$$

$$\sum_{j=1}^{M_i^Q}\left(\left[p_{ij}^S\right]^\wedge\right)_{3^*}^T\left(\left[p_{ij}^S\right]^\wedge\right)_{3^*}=\begin{bmatrix}\Gamma_{22}&-\Gamma_{12}&0\\-\Gamma_{12}&\Gamma_{11}&0\\0&0&0\end{bmatrix}=V_9 \tag{7.23}$$

With (7.18)–(7.23), $-\sum_{j=1}^{M_i^Q}\left[p_{ij}^S\right]^\wedge R_K^T\left(\Sigma_i^S\right)^{-1}R_K\left[p_{ij}^S\right]^\wedge$ is converted to (7.24). Then, (7.9) can be obtained by combining (7.24) and (7.17):

$$-\sum_{j=1}^{M_i^Q}\left[p_{ij}^S\right]^\wedge R_K^T\left(\Sigma_i^C\right)^{-1}R_K\left[p_{ij}^S\right]^\wedge=\begin{bmatrix}V_4\odot W&V_5\odot W&V_6\odot W\\V_5\odot W&V_7\odot W&V_8\odot W\\V_6\odot W&V_8\odot W&V_9\odot W\end{bmatrix}=\Omega_2 \tag{7.24}$$

By solving the incremental equation, the increment δx is acquired. Because $\delta x(1:3)$ is in the Lie algebra space, T_K is updated by $T_K\leftarrow T_K\boxplus\delta x$ after each iteration, where $\boxplus$ is the generalized sum operator [17]. The initial value of T_K is set as $\breve{T}_{K-1}\left(\breve{T}_{K-2}\right)^{-1}\breve{T}_{K-1}$, where $\breve{T}_{K-1}$ and $\breve{T}_{K-2}$ are the optimal poses obtained in frame $K-1$ and frame $K-2$ by the scan-to-map matching module. $\breve{T}_1$ and the initial value of T_2 are $\left(I_{3\times3},0_{3\times1}\right)$. To improve the iterative quality, data association is re-performed at regular iteration intervals. After the iteration process is finished, the estimated optimal pose $\breve{T}_K$ of frame K is obtained. Next, we calculate the distance of the LiDAR movement relative to the frame where the local maps are last updated. If their distance exceeds a preset threshold, the ground and non-ground point clouds of current frame will be downsampled and registered in the corresponding local maps, respectively. This completes the local map maintenance.

7.3.2 Fixed-Lag Smoothing

To further improve the accuracy of odometry estimation, a fixed-lag smoothing module is appended to the scan-to-map matching module. In reference [18], the eigenvalue of the covariance matrix for local feature points is minimized to refine the local consistency of the map. Inspired by this, we endeavor to maximize the likelihood of local point set for improving the consistency of odometry trajectory. Our module takes the estimated pose $\bar{T}_K$ and original point cloud as inputs, and optimizes a pose set of length L for more accurate pose output $\hat{T}_K$.

In the fixed-lag smoothing module, a voxel map is maintained within the sliding window. Firstly, the transformation $\bar{T}_K = \hat{T}_{K-1} \cdot \left(\bar{T}_{K-1}\right)^{-1} \cdot \bar{T}_K$ is calculated, where $\hat{T}_{K-1}$ is the optimized output of the smoothing module at frame $K-1$. When the current frame is selected into the smoothing module, its point cloud is downsampled and then registered into the voxel map with $\bar{T}_K$. A voxel in the map stores L point sets corresponding to L frames in the smoothing module as well as the marginalized point set moved out of the sliding window. It is noted that only point number, mean, and covariance of each aforementioned point set are kept. Herein, there is no need to search for correspondences as done in the scan-to-map module, and a voxel naturally constitutes a correspondence. To avoid the repeated calculation of distributions caused by multiple data associations, the point cloud is no longer adjusted after registration. Therefore, the states to be optimized become corrections $\Delta T_k = (\Delta R_k, \Delta t_k) \in SO(3) \times \mathcal{R}^3$, where k is the index of frames in the sliding window and $1 \le k \le L$. It is worth mentioning that not all of the voxels in the map are valuable. Only a voxel with enough points (excluding the marginalized points) is qualified for optimization. After all eligible voxels are determined, the objective function is constructed by maximizing the log-likelihood sum of points in different voxels as follows:

$$\max_{\Delta T_k, 1 \le k \le L} -\frac{1}{2} \sum_{i=1}^{N_f} w_i^f \sum_{k=1}^{L} \sum_{j=1}^{M_i^k} \left[e_{ij}^k(\Delta T_k) \right]^T \Sigma_i^{-1} \left[e_{ij}^k(\Delta T_k) \right] \tag{7.25}$$

where

$$e_{ij}^k(\Delta T_k) = \Delta R_k p_{ij}^k + \Delta t_k - \mu_i \tag{7.26}$$

$$\mu_i = \frac{1}{M_i^m + \sum_{k=1}^{L} M_i^k} \left[M_i^m \mu_i^m + \sum_{k=1}^{L} M_i^k \mu_i^k \right] \tag{7.27}$$

$$\Sigma_i = \frac{M_i^m \left[\Sigma_i^m + \mu_i^m \left(\mu_i^m\right)^T \right] + \sum_{k=1}^{L} \left[M_i^k \left(\Sigma_i^k + \mu_i^k \left(\mu_i^k\right)^T \right) \right]}{M_i^m + \sum_{k=1}^{L} M_i^k} - \mu_i \mu_i^T \tag{7.28}$$

$M_i^m, \mu_i^m, \Sigma_i^m$ refer to the point number, mean, and covariance of the marginalized point set in the ith voxel, respectively. $M_i^k, \mu_i^k, \Sigma_i^k$ are the point number, mean, and covariance of the point set from frame k in the ith voxel. μ_i and Σ_i denote mean and covariance of all points in the ith voxel, respectively. p_{ij}^k stands for the jth point of frame k in the ith voxel, and N_f represents the number of all eligible voxels. $w_i^f = \sigma_g \cdot d_{\mathcal{V}_i \mathcal{W}_i}$ is the weight for the ith voxel, where $\mathcal{W}_i$ and $\mathcal{V}_i$ refer to the sets of all points and non-marginalized points in this voxel, respectively. The geometry σ_g is calculated from $\mathcal{W}_i$ and will be set to zero when it is small. With (7.1) and (7.25), one can see that the likelihood of correspondence is extended from two frames to multiple frames. The purpose of the marginalized point is to provide a prior distribution for μ_i and Σ_i. Similar to the solution of (7.1), (7.25) is also solved by combining the two-step policy with Gauss-Newton algorithm.

In (7.25), one can observe that the weight w_i^f is independent of the index k, which means that the first two summations can be exchanged. If this occurs, (7.25) will become a form similar to the summation of (7.1) for L frames. Therefore, the incremental equation of (7.25) can be decomposed into L independent incremental equations with respect to different $\Delta T_k (1 \le k \le L)$. The incremental equation of (7.25) is described as follows and $H_k \delta x_k = g_k$ is the kth incremental equation:

$$
\begin{bmatrix}
H_1 & \cdots & 0 & \cdots & 0 \\
\vdots & \ddots & \vdots & \ddots & \vdots \\
0 & \cdots & H_k & \cdots & 0 \\
\vdots & \ddots & \vdots & \ddots & \vdots \\
0 & \cdots & 0 & \cdots & H_L
\end{bmatrix}
\begin{bmatrix}
\delta x_1 \\
\vdots \\
\delta x_k \\
\vdots \\
\delta x_L
\end{bmatrix}
=
\begin{bmatrix}
g_1 \\
\vdots \\
g_k \\
\vdots \\
g_L
\end{bmatrix}
\tag{7.29}
$$

where

$$
H_k = \sum_{i=1}^{N_f} w_i^f \sum_{j=1}^{M_i^k} \left(J_{ij}^k \right)^T \Sigma_i^{-1} J_{ij}^k
\tag{7.30}
$$

$$
g_k = -\sum_{i=1}^{N_f} w_i^f \sum_{j=1}^{M_i^k} \left(J_{ij}^k \right)^T \Sigma_i^{-1} e_{ij}^k
\tag{7.31}
$$

$$
J_{ij}^k = \frac{\partial e_{ij}^k (\Delta T_k)}{\partial \Delta T_k} = \left[-\Delta R_k \left[p_{ij}^k \right]^{\wedge} \quad I_{3\times3} \right]
\tag{7.32}
$$

For $\Delta T_k (1 \le k \le L)$, the previous optimized results of fixed-lag smoothing are used as the initial values of the iteration and the initial value of ΔT_L is set to be the same as ΔT_{L-1}. Considering the expansion of computation amount relative to the scan-to-map matching module, the number of iterations is reduced accordingly. It needs to be pointed out that we select a frame involved in the smoothing operation every other frame [19]. When current frame is not smoothed, $\bar{T}_K$ is directly regarded as the output $\hat{T}_K$ of this module, or else, $\hat{T}_K$ is given by

$$
\hat{T}_K = \Delta T_L^* \cdot \bar{T}_K
\tag{7.33}
$$

where ΔT_L^* is the locally optimal result corresponding to the Kth frame. After that, to maintain the fixed length of the sliding window, we marginalize the oldest frame in the map, and the point number, mean, and covariance of the oldest frame in the ith voxel are merged into M_i^m, μ_i^m, and Σ_i^m, where $i = 1, 2, \ldots, N_f$. Also, the oldest pose in the sliding window is removed. Note that the voxel map is initialized by registering and then marginalizing the point cloud of the first frame added into the smoothing module.

It is important to point out that the increment δx_k in (7.29) is dependent on H_k and g_k, which are only related to the points of the kth frame in the sliding window. Thus, it seems that there is no constraint among different ΔT_k. Actually, the constraints are reflected in (7.25) by the mean μ_i and covariance Σ_i of union of all point sets in a voxel, which explicitly incorporates the information of different ΔT_k. Thanks to the two-step policy that first updates the mean and covariance and then optimizes the poses, not only is the complexity of solving the optimization problem reduced, but also the constraints are maintained. Besides, different from BALM [18] that relies on the sparse feature points extracted from the front-end odometry module, our designed fixed-lag smoothing module is independent of the previous scan-to-map matching module. As a result, it can be attached to other LiDAR odometry as an independent unit to reduce the accumulated error of the pose estimation.

Finally, the outputs of scan-to-map matching and fixed-lag smoothing are combined by the strategy in [19]. For every frame, the pose increment relative to the previous frame is solved from the result of scan-to-map matching, and then it is cumulatively appended to the latest output of the fixed-lag smoothing. This integration result is taken as the final output $\mathcal{O}_K$ of our method whose output frequency is associated with that of scan-to-map matching regardless of the fixed-lag smoothing module.

7.4 EXPERIMENTS

The proposed method termed as HDLO is evaluated on the public KITTI dataset [20] and KITTI-360 dataset [21]. The KITTI dataset covers a variety of scenarios like urban streets, rural roads, and highways, where the ground truth trajectories for sequences 00–10 are provided. The point cloud for all sequences is collected by a car with a Velodyne HDL-64E laser scanner at a frequency of 10 Hz. As a successor of the KITTI dataset, KITTI-360 is a large-scale dataset, which adopts a similar data collection setup but covers wider suburb scenes with a driving distance of 73.7 km. Compared to KITTI, the trajectories of KITTI-360 are longer and irregular, which imposes challenge to localization.

Our method is implemented by C++ with OpenMP Library and executed on the robot operating system (ROS) [22] in a PC with an Intel Core i7-7700 CPU and 16 GB RAM. Similar to [23], we correct all the point cloud of the KITTI dataset with 0.195° in the pitch angle direction. In the following experiments, relative translation error (RTE) and relative rotation error (RRE) in the KITTI benchmark are adopted as the evaluation metrics of the odometry performance.

7.4.1 Accuracy Evaluation

In order to illustrate the performance of our HDLO method, we conduct a comparison with existing odometry and SLAM methods including LOAM [19], LO-Net-M [24], SuMa++ [25], LiTAMIN2 [10], SOFT-VO [26], and SOFT-SLAM [26]. The first two methods belong to LiDAR odometry based on point features and deep learning, respectively, the third and fourth methods are LiDAR SLAM with semantics and distribution, and the last two correspond to vision-based VO and SLAM. The accuracy results of the above methods are directly from their papers. Table 7.1 presents the comparison results in terms of RTE and RRE. The best results are labeled in bold. It can be seen that our HDLO without smoothing achieves better average RTE (0.52%) and RRE (0.20°/100 m). The introduction of fixed-lag smoothing further improves the accuracy of relative translation and rotation with average RTE (0.50%) and RRE (0.16°/100 m). By applying the fixed-lag smoothing module after the scan-to-map matching module, the average RTE and RRE are reduced by 3.8% and 20%, respectively. The conclusion is also supported by Figure 7.4, where the odometry trajectories of the proposed HDLO for sequences 00, 01, 02, 05, 07, and 10 of the KITTI dataset are provided.

We also test our method on the KITTI-360 dataset. The comparison results with A-LOAM [27] and F-LOAM [28] are presented in Table 7.2 including the trajectory distance for each sequence. The results also show the effectiveness of the proposed HDLO. The odometry trajectories of the proposed HDLO for sequences 00, 02, and 06 of the KITTI-360 dataset are illustrated in Figure 7.5.

7.4.2 Ablation Study

In order to further demonstrate the performance of the proposed HDLO, its three variants HDLO-I, HDLO-II, and HDLO-III are considered according to whether the fixed-lag smoothing module, the weights w_i and w_i^f in the optimization functions (7.1) and (7.25) are utilized. Table 7.3 presents the results of different variants averaged over sequences 00–10 of the KITTI dataset in terms of RTE and RRE. In comparison with HDLO-I, HDLO-II obviously improves the pose accuracy. Combining the results of HDLO-III and HDLO, we can see that the introduction of fixed-lag smoothing with or without the weight w_i^f can reduce the relative error. The HDLO method achieves the best result in RTE and RRE.

To evaluate the effect of size of the local map and the length of the sliding window on performance, a parameter analysis is conducted, and the results are shown in Tables 7.4 and 7.5, respectively. It is observed from Table 7.4 that the estimation error of scan-to-map matching module generally decreases with the increase in local map size. Because the threshold of adding a new frame to map is set to 3 m, when the size of local map exceeds 15, there is less overlapping between the newest and oldest frames. This implies that the oldest frame contributes a little to the registration performance, and it is meaningless to enlarge the size of local map. In this chapter, we set the size of local map to 15. For the fixed-lag smoothing module, with the increasing length of the sliding window, the constraint between two window frames far apart from each other often becomes weak, with less improvement in local consistency. According to the results in Table 7.5, the size of sliding window is chosen as 20, considering accuracy and efficiency.

TABLE 7.1　Accuracy Comparison of Different Methods for Sequences 00–10 on the KITTI Dataset in Terms of RTE and RRE

Sequence	LOAM		LO-Net-M		SuMa++		LiTAMIN2		SOFT-VO		SOFT-SLAM		HDLO w/o Smoothing		HDLO	
	RTE	RRE	RTE	RRE	RTE	RRE	RTE	RRE	RTE	RRE	RTE	RRE	RTE	RRE	RTE	RRE
00	0.78	-	0.78	0.42	0.64	0.22	0.7	0.28	0.63	0.24	0.66	0.22	0.55	0.22	**0.53**	**0.19**
01	1.43	-	1.42	0.40	1.60	0.46	2.1	0.46	0.96	0.18	0.96	0.18	**0.52**	0.16	0.65	**0.11**
02	0.92	-	1.01	0.45	1.00	0.37	0.98	0.32	0.74	0.20	1.36	0.23	0.58	0.19	**0.54**	**0.16**
03	0.86	-	0.73	0.59	0.67	0.46	0.96	0.48	0.70	**0.23**	0.70	**0.23**	0.66	0.42	**0.54**	0.30
04	0.71	-	0.56	0.54	**0.37**	0.26	1.05	0.52	0.50	0.18	0.50	0.18	0.45	0.10	0.53	**0.08**
05	0.57	-	0.62	0.35	0.40	0.20	0.45	0.25	0.47	0.20	0.43	0.17	0.31	0.18	**0.28**	**0.14**
06	0.65	-	0.55	0.33	0.46	0.21	0.59	0.34	0.38	0.18	0.41	0.14	0.34	0.14	**0.29**	**0.09**
07	0.63	-	0.56	0.45	0.34	0.19	0.44	0.32	0.36	0.23	0.36	0.24	**0.29**	**0.17**	0.31	0.17
08	1.12	-	1.08	0.43	1.10	0.35	0.95	0.29	**0.78**	**0.21**	**0.78**	**0.21**	0.88	0.24	0.80	**0.21**
09	0.77	-	0.77	0.38	0.47	0.23	0.69	0.40	0.74	0.17	0.59	0.18	0.53	0.18	**0.46**	**0.14**
10	0.79	-	0.92	0.41	0.66	0.28	0.80	0.47	0.68	0.26	0.68	0.26	0.64	0.23	**0.60**	**0.18**
Avg	0.84	-	0.82	0.43	0.70	0.29	0.88	0.38	0.63	0.21	0.68	0.20	0.52	0.20	**0.50**	**0.16**

RRE, relative rotation error (°/100 m); RTE, relative translation error (%).

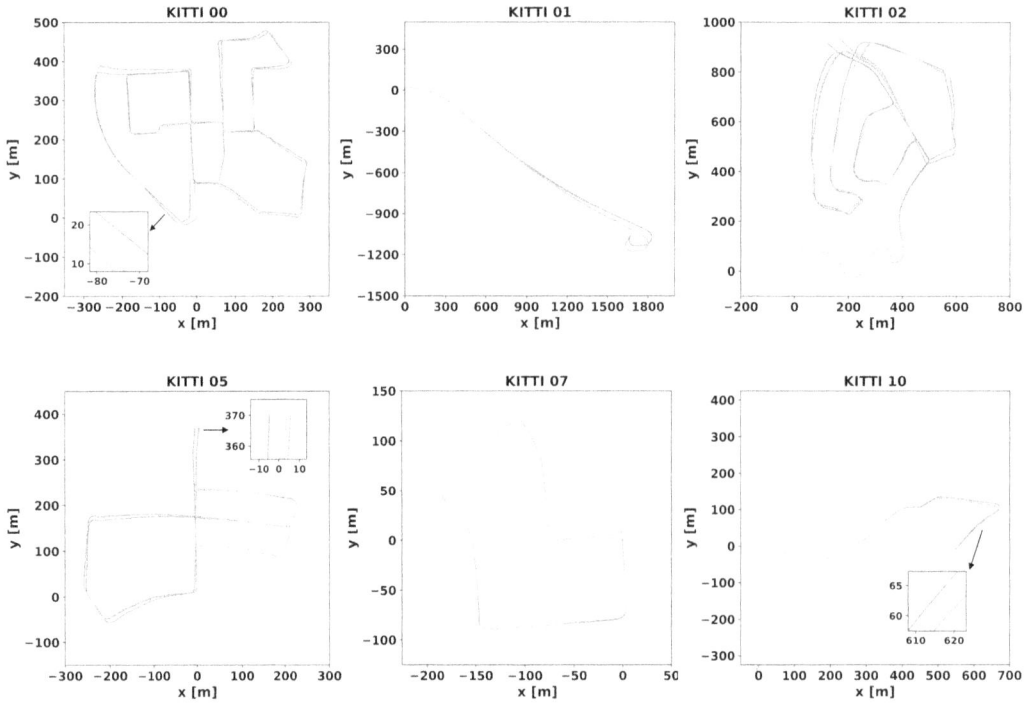

FIGURE 7.4   The trajectories of the proposed HDLO for sequences of the KITTI dataset.

TABLE 7.2   Accuracy Comparison of Different Methods on the KITTI-360 Dataset in Terms of RTE and RRE

	A-LOAM		F-LOAM		HDLO w/o Smoothing		HDLO	
Sequence	RTE	RRE	RTE	RRE	RTE	RRE	RTE	RRE
00 (8.39 km)	1.66	0.71	0.81	0.38	0.75	0.26	**0.57**	**0.14**
02 (15.33 km)	1.96	0.60	0.75	0.38	0.74	0.27	**0.68**	**0.22**
03 (1.38 km)	1.87	0.57	1.29	0.43	**0.52**	**0.17**	0.66	0.17
04 (9.96 km)	1.85	0.65	0.93	0.42	0.84	0.30	**0.75**	**0.25**
05 (4.68 km)	1.37	0.64	0.82	0.5	0.78	0.31	**0.67**	**0.28**
06 (7.97 km)	1.51	0.63	0.93	0.45	0.80	0.26	**0.69**	**0.20**
07 (4.89 km)	3.05	0.80	1.37	0.41	**1.10**	**0.27**	1.51	0.28
09 (10.57 km)	1.49	0.64	0.98	0.4	0.82	0.22	**0.77**	**0.19**
10 (3.34 km)	**0.77**	**0.23**	1.31	0.44	0.87	0.28	0.78	0.24

7.4.3 Efficiency Evaluation

In this section, the efficiency of our proposed method is demonstrated. The average running time per frame of the scan-to-map matching and fixed-lag smoothing modules on different sequences of KITTI is given in Table 7.6. One can infer that the scan-to-map matching and fixed-lag smoothing consume 53.87 and 43.47 ms on average, which shows the efficiency of our method. It can also be reflected in Figure 7.6, where the number of points before distribution extraction and that of distributions for data association on the sequence 00 of the KITTI dataset are provided. The average number of points and that of

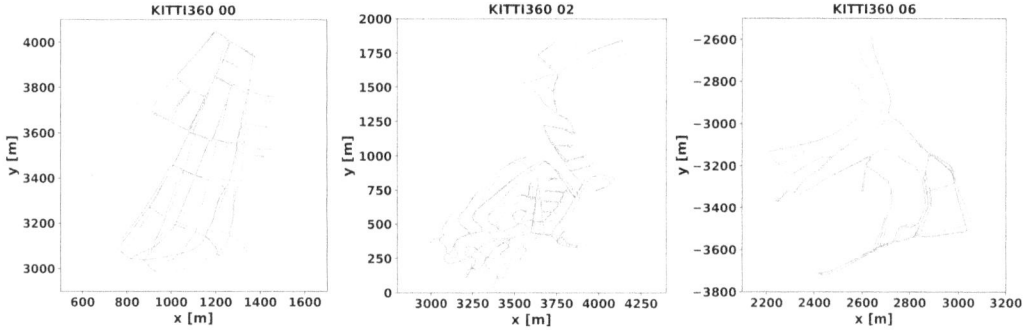

FIGURE 7.5   The trajectories of the proposed HDLO.

TABLE 7.3   Accuracy Comparison of Different Variants of Our Method Averaged over Sequences 00–10 of the KITTI Dataset

Method	Weight in Scan-to-Map Matching	Fixed-Lag Smoothing	Weight in Fixed-Lag Smoothing	RTE (%)	RRE (°/100 m)
HDLO-I	×	×	×	0.57	0.21
HDLO-II	√	×	×	0.52	0.20
HDLO-III	√	√	×	0.52	**0.16**
HDLO	√	√	√	**0.50**	**0.16**

TABLE 7.4   The Average Accuracy of HDLO w/o Smoothing for Different Sizes of the Local Map on Sequences 00–10 of the KITTI Dataset

	Local Map Size				
	5	10	15	20	25
RTE	0.58	0.53	**0.52**	0.52	0.52
RRE	0.24	0.21	**0.20**	0.20	0.20

TABLE 7.5   The Average Accuracy of HDLO for Different Sizes of the Sliding Window on Sequences 00–10 of the KITTI Dataset

	Sliding Window Size		
	10	20	30
RTE	0.51	**0.50**	0.50
RRE	0.17	**0.16**	0.16

TABLE 7.6   Average Running Time (ms) per Frame of Two Modules on Different Sequences of KITTI Dataset

	00	01	02	03	04	05	06	07	08	09	10
Scan-to-Map Matching	49.51	59.04	50.90	54.83	61.23	51.29	60.67	47.75	54.12	54.51	48.66
Fixed-Lag Smoothing	35.69	52.70	37.72	48.32	49.54	39.86	59.58	34.02	43.36	43.64	33.7

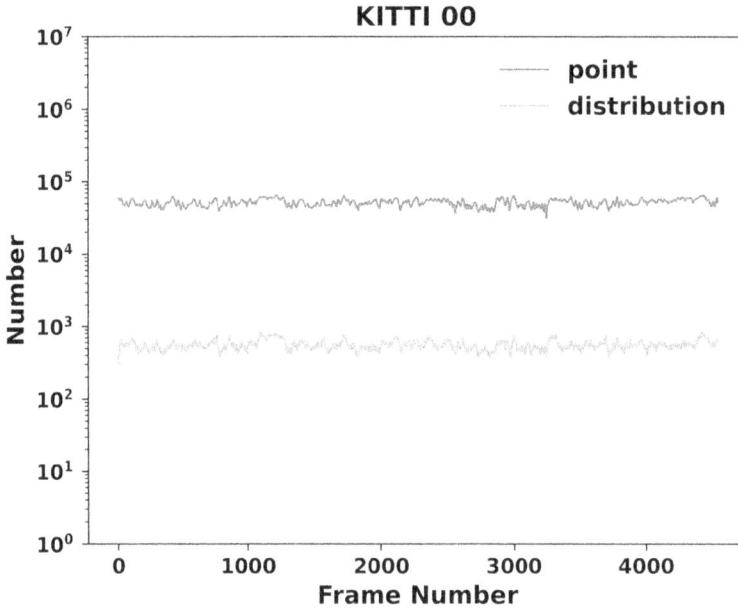

FIGURE 7.6 The comparison between the number of points before distribution extraction and that of distributions for data association in the scan-to-map matching module. The x-axis describes the frame number, and the y-axis is expressed in logarithmic coordinates.

distributions are 51650 and 558, respectively. Compared to the former, the number of distributions for data association is reduced in an order of two.

7.4.4 Verification in an Outdoor Scenario

We also test the proposed method on an outdoor scenario. The experiment employs a Robosense RS-Ruby laser scanner mounted on a mobile robot. In this experiment, the robot is controlled to move approximately 570 m and stop motion when returning to the neighborhood of the start position. The mapping result and the estimated trajectory for this outdoor scenario are depicted in Figure 7.7. Without the pose ground truth, we employ ICP to solve the relative translation between the start position and the end position, which is regarded as the ground truth. Then, the mean relative translation drift is used as the evaluation metric, where it is calculated by the error between the estimated translation and ground truth averaged over the total trajectory. Table 7.7 presents the result of HDLO on the outdoor scenario, where d_x, d_y, and d_z are translation drifts in the x, y, and z directions, respectively, and d_{mr} denotes the mean relative translation drift.

7.4.5 Generalization Evaluation of Fixed-Lag Smoothing

In the proposed method, fixed-lag smoothing provides a more accurate estimation result than scan-to-map matching. Note that the proposed smoothing module takes original point cloud as the input, and it is independent of the matching module. Therefore, this module can be directly attached to existing LiDAR odometry methods. Table 7.8 presents the results of A-LOAM [27] with or without our fixed-lag smoothing on the sequences

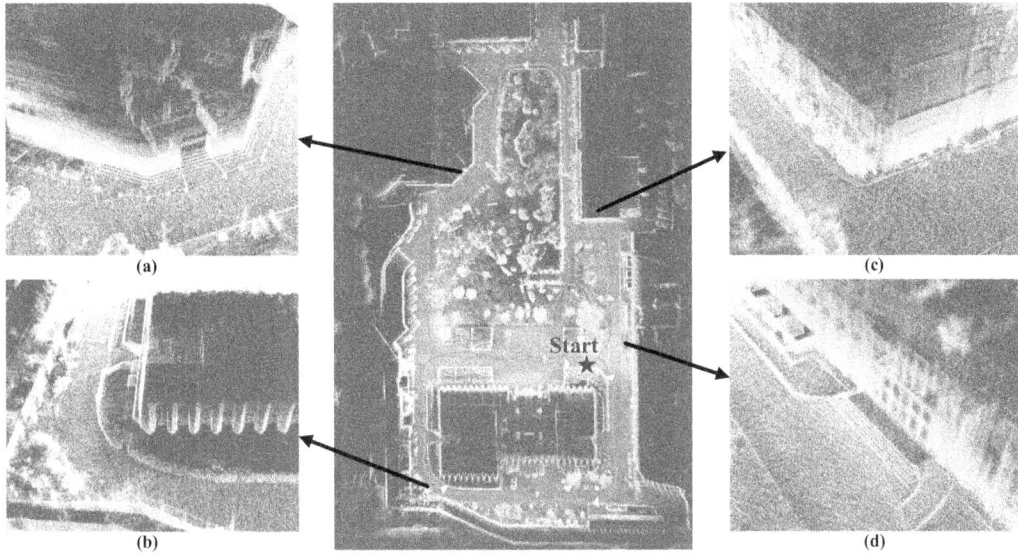

FIGURE 7.7   The mapping and the estimated trajectory for the outdoor scenario. (a)–(d) represent the details of four selected local regions in the point cloud map.

TABLE 7.7   Result of HDLO on the Outdoor Scenario

d_x	d_y	d_z	d_{mr}
0.35 m	0.17 m	0.076 m	0.07%

TABLE 7.8   Accuracy Comparison of A-LOAM with and without our Fix-Lag Smoothing Module on the KITTI Dataset

Method		00	01	02	03	04	05	06	07	08	09	10	Average
A-LOAM	RTE	0.76	0.75	4.17	0.90	0.54	0.54	0.40	0.44	1.02	0.81	1.02	1.03
	RRE	0.29	0.12	1.36	0.23	0.24	0.32	0.13	0.30	0.27	0.24	0.35	0.35
A-LOAM+ Smoothing	RTE	0.52	0.77	3.99	0.55	0.53	0.28	0.28	0.3	0.81	0.45	0.63	0.83
	RRE	0.19	0.12	1.17	0.3	0.08	0.14	0.09	0.17	0.21	0.13	0.18	0.25

00–10 of KITTI dataset. Although the A-LOAM is a LiDAR odometry based on the feature points, our smoothing module can contribute to the improvement of estimation accuracy, which proves the universality of our smoothing module.

7.5 CONCLUSION

This chapter presents a hierarchical LiDAR odometry based on maximum likelihood estimation with tightly associated distributions. A matching method with distribution-to-distribution correspondence is designed, which takes advantage of both source and target point cloud. With this matching method, a scan-to-map matching module is constructed for pose estimation at the low level. Furthermore, we extend the matching method to a fixed-lag smoothing solution by associating point sets among multiple frames in a sliding window. Accordingly, the accuracy of pose estimation is further improved. The experimental results demonstrate the effectiveness and advantage of our proposed method.

REFERENCES

[1] Biber, P., & Straßer, W. (2003). The normal distributions transform: A new approach to laser scan matching. In *Proceedings of the IEEE/RSJ International Conference on Intelligent Robots and Systems*, Las Vegas, NV, USA (pp. 2743–2748).

[2] Magnusson, M., Lilienthal, A., & Duckett, T. (2007). Scan registration for autonomous mining vehicles using 3D-NDT. *Journal of Field Robotics*, 24(10), 803–827.

[3] Segal, A., Haehnel, D., & Thrun, S. (2009). Generalized-ICP. In *Robotics: Science and Systems*, Seattle, USA (vol. 2, no. 4, p. 435)

[4] Yue, J., Wen, W., Han, J., & Hsu, L. T. (2021). 3D point clouds data super resolution-aided LiDAR odometry for vehicular positioning in urban canyons. *IEEE Transactions on Vehicular Technology*, 70(5), 4098–4112.

[5] Takeuchi, E., & Tsubouchi, T. (2006). A 3-D scan matching using improved 3-D normal distributions transform for mobile robotic mapping. In *Proceedings of the IEEE/RSJ International Conference on Intelligent Robots and Systems*, Beijing, China (pp. 3068–3073).

[6] Magnusson, M., Nuchter, A., Lorken, C., Lilienthal, A. J., & Hertzberg, J. (2009). Evaluation of 3D registration reliability and speed-A comparison of ICP and NDT. In *Proceedings of the IEEE International Conference on Robotics and Automation*, Kobe, Japan (pp. 3907–3912).

[7] Koide, K., Yokozuka, M., Oishi, S., & Banno, A. (2021). Voxelized GICP for fast and accurate 3D point cloud registration. In *Proceedings of the IEEE International Conference on Robotics and Automation*, Xi'an, China (pp. 11054–11059).

[8] Vlaminck, M., Luong, H., & Philips, W. (2018). Surface-based GICP. In *15th Conference on Computer and Robot Vision*, Toronto, ON, Canada (pp. 262–268).

[9] Yokozuka, M., Koide, K., Oishi, S., & Banno, A. (2020). LiTAMIN: LiDAR-based tracking and mapping by stabilized ICP for geometry approximation with normal distributions. In *Proceedings of the IEEE/RSJ International Conference on Intelligent Robots and Systems*, Las Vegas, NV, USA (pp. 5143–5150).

[10] Yokozuka, M., Koide, K., Oishi, S., & Banno, A. (2021). LiTAMIN2: Ultra Light LiDAR-based SLAM using geometric approximation applied with KL-divergence. In *Proceedings of the IEEE International Conference on Robotics and Automation*, Xi'an, China (pp. 11619–11625).

[11] Stoyanov, T., Magnusson, M., Andreasson, H., & Lilienthal, A. J. (2012). Fast and accurate scan registration through minimization of the distance between compact 3D NDT representations. *The International Journal of Robotics Research*, 31(12), 1377–1393.

[12] Tabib, W., O'Meadhra, C., & Michael, N. (2018). On-Manifold GMM registration. *IEEE Robotics and Automation Letters*, 3(4), 3805–3812.

[13] Eckart, B., Kim, K., & Kautz, J. (2018). HGMR: Hierarchical Gaussian mixtures for adaptive 3D registration. In *Proceedings of the European Conference on Computer Vision*, Munich, Germany (pp. 730–746).

[14] Zermas, D., Izzat, I., & Papanikolopoulos, N. (2017). Fast segmentation of 3D point clouds: A paradigm on LiDAR data for autonomous vehicle applications. In *Proceedings of the IEEE International Conference on Robotics and Automation*, Singapore (pp. 5067–5073).

[15] Hackel, T., Wegner, J. D., & Schindler, K. (2016). Fast semantic segmentation of 3D point clouds with strongly varying density. *ISPRS Annals of the Photogrammetry, Remote Sensing and Spatial Information Sciences*, 3, 177–184.

[16] Herdin, M., Czink, N., Ozcelik, H., & Bonek, E. (2005). Correlation matrix distance, a meaningful measure for evaluation of non-stationary MIMO channels. In *IEEE Vehicular Technology Conference*, Stockholm, Sweden (pp. 136–140).

[17] Hertzberg, C., Wagner, R., Frese, U., & Schröder, L. (2013). Integrating generic sensor fusion algorithms with sound state representations through encapsulation of manifolds. *Information Fusion*, 14(1), 57–77.

[18] Liu, Z., & Zhang, F. (2021). BALM: Bundle adjustment for Lidar mapping. *IEEE Robotics and Automation Letters*, 6(2), 3184–3191.

[19] Zhang, J., & Singh, S. (2014). LOAM: Lidar Odometry and mapping in real-time. *Robotics: Science and Systems*, 2(9), 1–9.

[20] Geiger, A., Lenz, P., & Urtasun, R. (2012). Are we ready for autonomous driving? The KITTI vision benchmark suite. In *Proceedings of IEEE Conference on Computer Vision and Pattern Recognition*, Providence, RI, USA (pp. 3354–3361).

[21] Liao, Y., Xie, J., & Geiger, A. (2022). KITTI-360: A Novel dataset and benchmarks for urban scene understanding in 2D and 3D. *IEEE Transactions on Pattern Analysis and Machine Intelligence*, 45(3), 3292–3310.

[22] Quigley, M., Conley, K., Gerkey, B., Faust, J., Foote, T., Leibs, J., Berger, E., Wheeler, R., & Ng, A. Y. (2009). ROS: An open-source Robot Operating System. In *ICRA Workshop on Open Source Software* (vol. 3, no. 3.2, pp. 1–6).

[23] Deschaud, J. E. (2018). IMLS-SLAM: Scan-to-model matching based on 3D data. In *Proceedings of the IEEE International Conference on Robotics and Automation*, Brisbane, QLD, Australia (pp. 2480–2485).

[24] Li, Q., Chen, S., Wang, C., Li, X., Wen, C., Cheng, M., & Li, J. (2019). LO-Net: Deep real-time Lidar odometry. In *Proceedings of the IEEE/CVF Conference on Computer Vision and Pattern Recognition*, Long Beach, CA, USA (pp. 8465–8474).

[25] Chen, X., Milioto, A., Palazzolo, E., Giguere, P., Behley, J., & Stachniss, C. (2019). SuMa++: Efficient LiDAR-based Semantic SLAM. In *Proceedings of the IEEE/RSJ International Conference on Intelligent Robots and Systems*, Macau, China (pp. 4530–4537).

[26] Cvišić, I., Ćesić, J., Marković, I., & Petrović, I. (2018). SOFT-SLAM: Computationally efficient stereo visual SLAM for autonomous UAVs. *Journal of Field Robotics*, 35(4), 578–595.

[27] Qin, T., & Cao, S. (2015). A-LOAM. Retrieved from https://github.com/HKUST-Aerial-Robotics/A-LOAM

[28] Wang, H., Wang, C., Chen, C. L., & Xie, L. (2021). F-LOAM: Fast LiDAR odometry and mapping. In *Proceedings of the IEEE/RSJ International Conference on Intelligent Robots and Systems*, Prague, Czech Republic (pp. 4390–4396).

CHAPTER 8

Hierarchical Distribution-Based Tightly Coupled LiDAR Inertial Odometry

8.1 INTRODUCTION

For LiDAR inertial odometry (LIO), it is typically divided into two categories: loosely coupled and tightly coupled LIO. The former takes IMU as assistance of LO. By contrast, the tightly coupled LIO directly builds joint constraints with the measurements from different sensors instead of odometry results. It becomes a focus of current research due to good localization performance. Considering that the distribution can fully exploit point cloud information, distribution-based tightly coupled LIO is attractive. A challenge arises when coupling the distribution-based constraint of point cloud and constraint of inertial measurements. The distribution-based constraint employs the form of Mahalanobis distance from the point to distribution whose observation is related to the decomposed component of the distribution covariance. By summarizing the squares of the observation residual norms constructed from the measurements of different sensors, joint constraints are built and poses of LiDAR are optimized. However, LiDAR measurement noises inevitably bring in the uncertainty of observation residual. The uncertainty also involves the component of distribution covariance. When weighting the point cloud constraint with the inverse of residual uncertainty, their corresponding components of distribution covariance are neutralized. This leads to the degeneration of constraint, which will reduce the accuracy. It is also worth mentioning that the existing tightly coupled methods fuse the measurements from LiDAR and IMU by either filtering [1] or graph optimization [2]. The former usually focuses on the newest frame, which enables efficient odometry output. The latter can

 DOI: 10.1201/9781003643630-8

optimize multiple frames simultaneously and more precise but slower localization results are expected. Their integration deserves further investigation.

For the point-to-distribution distance constraint degeneration problem, it is caused by the component related to the covariance in the corresponding residual, which essentially plays a standardization role of residual. This component contains eigenvalue elements and eigenvector matrix. Because the function of the eigenvalue elements is to weight residual parts in different directions, we design a special loss function to exclude the eigenvalue elements out of the observation. With this rectification, the neutralization of components related to the covariance can be eliminated and then the constraint degeneration problem is solved. To exploit the advantages of filtering and graph optimization, this chapter combines them in a hierarchical framework. In the low level, the point-to-distribution distance constraints of the latest frame point cloud are joint optimized with the prior state constraints from IMU based on the iterated extended Kalman filter (IEKF), achieving an efficient but relatively coarse odometry estimation. The above result is then transferred to the high-level factor graph optimization for further refinement. By fusing the results from these two levels, a fast and precise odometry estimation is obtained.

In this chapter, a hierarchical tightly coupled LIO based on distribution is proposed. The main contributions are twofold. Firstly, the degeneration of point cloud constraints in distribution-based LIO is solved. According to the standardization function of components related to the covariance of the distribution in residual, the uncertainty of residual derived from the LiDAR measurement noise is rectified by a special loss function. Thus, the neutralization of components related to the covariance is eliminated and the anti-degeneration point-to-distribution constraints are obtained, which are jointly optimized with the IMU constraints for odometry estimation. It enables the effective coupling of LiDAR and IMU measurements without odometry precision deterioration. Secondly, a hierarchical distribution-based LIO framework is designed, combining a low-level tightly coupled LIO based on IEKF and a high-level tightly coupled factor graph optimization. The former is utilized to provide a fast state estimation by taking the advantage of only optimizing the latest frame in filtering, and the latter integrates a prior factor, IMU pre-integration factors, and LiDAR observation constraints among multiple frames for more precise estimation.

The rest of the chapter is organized as follows. Section 8.2 introduces the related work. In Section 8.3, the proposed hierarchical tightly coupled LIO framework is described in detail. The experiment results are presented in Section 8.4, and Section 8.5 concludes this chapter.

8.2 REVIEW OF LIDAR INERTIAL ODOMETRY/SLAM

The fusion of LiDAR and IMU provides an effective solution. According to their fusion way, the LIO or SLAM is generally divided into loosely coupled and tightly coupled types.

A popular loosely coupled implementation is to utilize IMU as aid of LiDAR odometry, where the inertial measurement is used to correct the motion distortion of LiDAR scan and provide prior information for point cloud registration. In [3], IMU is adopted to remove the rotational and nonlinear motion distortion, which enables the proposed odometry methods to satisfy the smooth and continuous motion assumption. Chen et al. [4] presented a direct LiDAR odometry, which accepts a rotational prior from the gyroscopic measurement pre-integration [5] to improve accuracy during aggressive rotational motions. Palieri et al. [6] incorporated a sensor prior module into the LiDAR scan matching, where this module is applied to provide a prior pose according to measurements of IMU. It can also be extended with other prior sources such as visual measurements. In their follow-up work [7], IMU measurements are additionally utilized for motion distortion correction. Besides, with the assumption of LiDAR uniform motion, the roll and pitch can be directly solved from the linear acceleration measurements of IMU. On this basis, [8] simplifies the pose estimation from 6-DOF (degrees of freedom) to 4-DOF, and [9] reduces the search dimensions of scan-to-submap matching for finding loop closure constraints. There are other researches concentrating on the fusion of odometry results derived from different sensors. Hening et al. [10] directly formulated the observation with the velocity and position estimations from LiDAR SLAM and GPS under the adaptive extended Kalman filter, where the observation is used to correct the prediction derived from high-frequency IMU measurements. However, the estimation failure of one sensor source may cause the system to deteriorate and even diverge, affecting its precision and robustness.

Different from the odometry constraints in loosely coupled methods, the tightly coupled solution concerns the observation constraints constructed from the raw measurements of different sensors, which are added for joint optimization. On one hand, the information from IMU prediction and LiDAR measurements are simultaneously considered, which compensates for the missing constraints of a single sensor in some under-constraint situations such as long corridor. This is beneficial to improve the aforementioned deterioration problem in loosely coupled methods, increasing system robustness. On the other hand, exploiting more raw information also decreases the uncertainties of pose estimation [11], which brings in the improvement of precision. Tightly coupled LIO can be implemented by filtering or graph optimization. In the filtering-based methods [1,12–14], point features are employed to build the observation equations of IEKF in the form of point-to-line or point-to-plane distance. With the plane features, Hesch et al. constructed the orientation and distance constraints based on line-to-plane correspondences to update the state estimations in EKF [15]. In these methods, only the latest pose is optimized, which will affect the odometry precision. In order to deal with the accumulative error better, the approaches based on pose graph appear, where the error is spread to the different poses via multi-frame simultaneous optimization at the expense of higher computation. In [16–18], the pose graph optimization is utilized to integrate IMU pre-integration [5] and LiDAR odometry constraints between neighboring frames. The LiDAR odometry constraints in pose graph are obtained according to different registration methods including GICP [17], ICP variant [16], and occupancy grid-based matching [18]. A problem is that incorrect odometry estimation will result in erroneous constraints, whose negative effects cannot be eliminated once added to the pose graph.

Besides, the factor graphs are employed for fusion by replacing the LiDAR odometry constraints with point-to-line and point-to-plane constraints [2]. With more observation constraints, this solution generally attains higher precision than the pose graph-based methods. However, more constraints bring in heavier computation load. To solve this problem, the factor graphs are implemented based on keyframes within a sliding window [19,20]. It is noted that filtering and graph optimization have their respective strengths and weaknesses and their combination can promote the improvement of performance. In this chapter, we propose a tightly coupled LIO based on distribution, where an EKF-based LIO and a factor graph optimization are organized in a hierarchical form, obtaining good efficiency and precision.

8.3 HIERARCHICAL DISTRIBUTION-BASED TIGHTLY COUPLED LIDAR INERTIAL ODOMETRY

The proposed tightly coupled hierarchical LIO is composed of a low-level LIO based on EKF and a high-level factor graph optimization. As shown in Figure 8.1, at the low level, the recent IMU measurements are firstly utilized to predict the prior state and correct the motion distortion in LiDAR scan. Then, with the undistorted point cloud $^{I_k}\mathcal{P}$, distributions are extracted and distribution correspondences are found by data association. Considering the uncertainty of LiDAR measurement noises, non-degenerate residuals are constructed and transferred to IEKF for updating the prior prediction, obtaining a fast state estimation $\hat{X}_k$. To further refine the estimation and enhance the local consistency of the trajectory, a factor graph of length L is employed in the high level for joint optimization of multiple

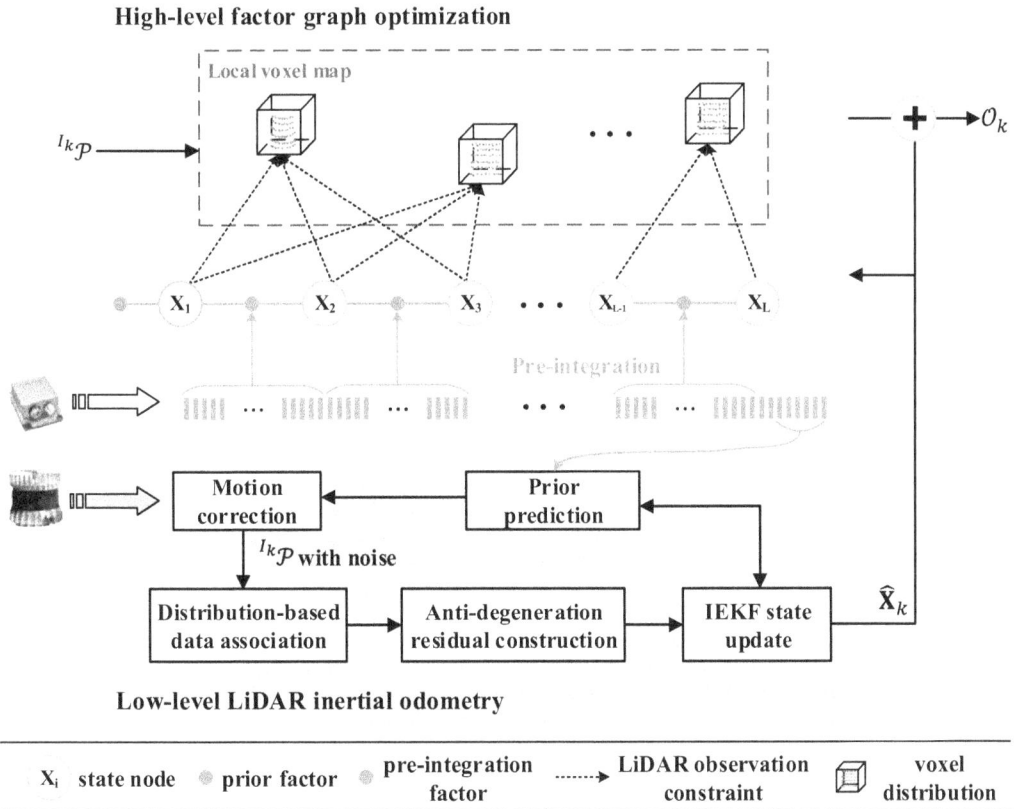

FIGURE 8.1   Pipeline of the hierarchical tightly coupled LiDAR inertial odometry.

window frames. The factor graph contains a prior factor, IMU pre-integration factors, and LiDAR observation constraints based on local voxel map. The high-fidelity result by graph optimization is integrated with $\hat{X}_k$ as the final output $\mathcal{O}_k$ of the proposed method.

8.3.1 Problem Statement

Our aim is to estimate the state of robot frame relative to the world frame with the LiDAR and IMU measurements based on the low-level EKF and high-level graph optimization. With the prior prediction of IMU, the LiDAR measurements are utilized to correct the predicted state in the EKF update. After distributions are extracted, data association between current frame and local map is executed to obtain correspondences and point cloud residuals are built. The distribution and data association are from Chapter 7. Particularly, when building point cloud residuals for coupling, this chapter additionally takes the LiDAR noise into account without degenerating the constraints. And then the state estimation from EKF is transferred to the factor graph optimization (prior factor, IMU pre-integration factor, and LiDAR observation constraints) for further refinement.

We denote LiDAR and IMU frames at time t_k as L_k and I_k, respectively. The robot frame is defined to coincide with the IMU frame, and the first IMU frame is taken as the world frame W. The transformation from LiDAR to IMU is denoted as ${}^I T_L = \left({}^I R_L, {}^I t_L \right)$, where ${}^I R_L$ and ${}^I t_L$ refer to rotation and translation. The state X_k at time t_k is defined as

$$X_k = \begin{bmatrix} {}^W p_{I_k}{}^T & {}^W v_{I_k}{}^T & {}^W R_{I_k}{}^T & b_{I_k}^{a\,T} & b_{I_k}^{\omega T} \end{bmatrix}^T \in \mathbb{R}^6 \times SO(3) \times \mathbb{R}^6 \quad (8.1)$$

where ${}^W R_{I_k}$ represents the rotation from I_k to the world coordinate system W. ${}^W p_{I_k}$ and ${}^W v_{I_k}$ refer to the position and velocity of IMU in W. $b_{I_k}^a$ and $b_{I_k}^\omega$ represent the bias of IMU acceleration and angular velocity measurements, respectively. $SO(3)$ refers to the special orthogonal group in three-dimensional space.

As EKF cannot be applied to the manifold space, we focus on error state $\tilde{X}_k$, which is defined below:

$$\tilde{X}_k = X_k \boxminus \breve{X}_k = \begin{bmatrix} {}^W \tilde{p}_k{}^T & {}^W \tilde{v}_k{}^T & \tilde{\theta}_k{}^T & \tilde{b}_k^{aT} & \tilde{b}_k^{\omega T} \end{bmatrix}^T \in \mathbb{R}^{15} \quad (8.2)$$

where $\breve{X}_k$ is the estimation of X_k and $\boxminus$ represents the general minus operation. $\tilde{\theta}_k = {}^W R_{I_k} \boxminus {}^W \breve{R}_{I_k} = \log\left(\left({}^W \breve{R}_{I_k} \right)^T \cdot {}^W R_{I_k} \right)$ is the rotation error. ${}^W \tilde{p}_k$ and ${}^W \tilde{v}_k$ refer to the errors of position and velocity. $\tilde{b}_k^a$ and $\tilde{b}_k^\omega$ represent the bias errors for IMU acceleration and angular velocity measurements. Denote the optimal estimation and the covariance of corresponding error state estimation at time t_{k-1} as $\hat{X}_{k-1}$ and $\hat{P}_{k-1}$, respectively. With the IMU measurements at $(t_{k-1}, t_k]$, we obtain the predicted state $\bar{X}_k$ and covariance $\bar{P}_k$ at t_k according to the kinematics of IMU [21].

Define the kth frame point cloud in LiDAR frame from t_{k-1} to t_k as ${}^{L_k}\mathcal{P}$. A radius filtering is first performed on ${}^{L_k}\mathcal{P}$ to remove the invalid points around the LiDAR center. Then the preprocessed point cloud is transformed to the IMU frame and corrected [17], and we have the undistorted point cloud $^{I_k}\mathcal{P}$. Using the distribution extraction and data association in Chapter 7, a set of distribution correspondences $\left\{(\mathcal{S}_i,\mathcal{T}_i)\,|\,\mathcal{S}_i \sim \mathcal{N}\left(^{I_k}\mu_i^S,\,^{I_k}\Sigma_i^S\right),\mathcal{T}_i \sim \mathcal{N}\left(^W\mu_i^T,\,^W\Sigma_i^T\right),1 \le i \le N\right\}$ is obtained, where $\mathcal{S}_i$ and $\mathcal{T}_i$ are the ith source distribution and corresponding target distribution, respectively. N is the number of correspondences. Note that the ground and non-ground points are not differentiated to improve the completeness of the distributions nearby the ground plane, which decreases the drift in the z-axis. Denote the union of $\mathcal{Q}_i$ and $\mathcal{T}_i$ as $\mathcal{C}_i \sim \mathcal{N}\left(^W\mu_i^C,\,^W\Sigma_i^C\right)$ for constructing the distance constraints from point to distribution, where $\mathcal{Q}_i \sim \mathcal{N}\left(^W\mu_i^Q,\,^W\Sigma_i^Q\right)$ refers to the transformed distribution of $\mathcal{S}_i$, $^W\mu_i^Q = {}^WR_{I_k} \cdot {}^{I_k}\mu_i^S + {}^Wp_{I_k}$, $^W\Sigma_i^Q = {}^WR_{I_k} \cdot {}^{I_k}\Sigma_i^S \cdot \left(^WR_{I_k}\right)^T$.

The objective of this chapter is as follows. Given the point cloud $^{I_k}\mathcal{P}$ and IMU measurements within $(t_{k-1},t_k]$, construct the constraint of IMU prediction prior and point cloud constraint under LiDAR measurement noise, then obtain a coarse estimation of state X_k based on EKF framework, on this basis, optimize multi-frame poses in a sliding window with a factor graph, such that an efficient and precise odometry is achieved.

8.3.2 Low-Level LiDAR Inertial Odometry

The prior state constraint of IMU is first constructed as follows:

$$\left\|X_k \boxminus \bar{X}_k\right\|^2_{\bar{P}_k^{-1}} = \left\|\bar{X}_k \boxplus \tilde{X}_k \boxminus \bar{X}_k\right\|^2_{\bar{P}_k^{-1}} \tag{8.3}$$

where $\|a\|^2_G = a^T Ga$ refers to Mahalanobis norm.

For LiDAR, the source-target distribution correspondences $(\mathcal{S}_i,\mathcal{T}_i)$ as well as the union $\mathcal{C}_i$ are acquired after data association, where $1 \le i \le N$. Then the decomposition of point-to-distribution distance is utilized to build observation equation. For the jth point $^{I_k}\mathbf{p}_{ij}^S$ in the source distribution $\mathcal{S}_i$, its observation equation is constructed as below:

$$
\begin{aligned}
z_{ij} &= \Lambda_i^{-\frac{1}{2}} U_i^T \left[^WR_{I_k}\left(^IR_L\left(^{I_k}\mathbf{p}_{ij}^S + n\right) + {}^It_L\right) + {}^Wp_{I_k} - {}^W\mu_i^C\right] \\[6pt]
&= \Lambda_i^{-\frac{1}{2}} U_i^T \left[^WR_{I_k}\left(^{I_k}\mathbf{p}_{ij}^S + {}^IR_L n\right) + {}^Wp_{I_k} - {}^W\mu_i^C\right] \\[6pt]
&= \Lambda_i^{-\frac{1}{2}} U_i^T e_{ij} + v_i
\end{aligned}
\tag{8.4}
$$

where $v_i = \Lambda_i^{-\frac{1}{2}} U_i^T {}^WR_{I_k}{}^IR_L n = \Lambda_i^{-\frac{1}{2}} U_i^T {}^WR_{L_k} n$ and $e_{ij} = {}^WR_{I_k}{}^{I_k}\mathbf{p}_{ij}^S + {}^Wp_{I_k} - {}^W\mu_i^C$. n is the measurement noise of LiDAR, and it obeys an isotropic Gaussian distribution $\mathcal{N}(0,\Sigma_n)$, $\Sigma_n = \sigma I_{3\times3}$. It can be inferred that v_i follows the Gaussian distribution $\mathcal{N}\left(0, L_{ij}\Sigma_n L_{ij}^T\right)$, where $L_{ij} = \Lambda_i^{-\frac{1}{2}} U_i^T {}^WR_{L_k}$, Λ_i and U_i are eigenvalue matrix and eigen vector matrix of $^W\Sigma_i^C$, respectively. According to the characteristics of eigenvalue decomposition of real symmetric matrix, Λ_i is a diagonal matrix and U_i is an orthogonal matrix. Thus, ${}^W\Sigma_i^{\mathcal{C}}$, Λ_i, and U_i satisfy

$${}^W\Sigma_i^{\mathcal{C}} = U_i\Lambda_i U_i^T = U_i\Lambda_i^{\frac{1}{2}}\Lambda_i^{\frac{1}{2}}U_i^T = \left(U_i\Lambda_i^{-\frac{1}{2}}\Lambda_i^{-\frac{1}{2}}U_i^T\right)^{-1}, \text{ in which } (\cdot)^{\frac{1}{2}} \text{ and } (\cdot)^{-\frac{1}{2}} \text{ represent diagonal}$$

element-wise operations.

The observation z_{ij} in (8.4) is expected to be zero vector. Then the residual for point ${}^{L_k}\mathbf{p}_{ij}^{\mathcal{S}}$ is defined as $h_{ij} = \Lambda_i^{-\frac{1}{2}}U_i^T e_{ij}$. If the noise is not considered, the constraint is expressed by $\|h_{ij}\|^2 = h_{ij}^T h_{ij} = e_{ij}^T U_i \Lambda_i^{-\frac{1}{2}}\Lambda_i^{-\frac{1}{2}}U_i^T e_{ij} = e_{ij}^T\left({}^W\Sigma_i^{\mathcal{C}}\right)^{-1} e_{ij}$, which is actually a form of point-to-distribution distance relevant to the covariance ${}^W\Sigma_i^{\mathcal{C}}$. It is noticed that $\Lambda_i^{-\frac{1}{2}}U_i^T$ acts as a standardization or PCA whitening, where U_i^T projects the error vector e_{ij} to the principal component directions of ${}^W\Sigma_i^{\mathcal{C}}$ and $\Lambda_i^{-\frac{1}{2}}$ is utilized to normalize them. The aforementioned conclusion is got without noise. In practice, the LiDAR noise is inevitable. Then, the point-to-distribution distance constraint becomes

$$\|h_{ij}\|^2_{\left(L_{ij}\Sigma_n L_{ij}^T\right)^{-1}}$$

$$= \left(h_{ij}\right)^T \left(L_{ij}\Sigma_n L_{ij}^T\right)^{-1} h_{ij}$$

$$= \left(\Lambda_i^{-\frac{1}{2}}U_i^T e_{ij}\right)^T \left[\Lambda_i^{-\frac{1}{2}}U_i^T\,{}^WR_{L_k}\Sigma_n\left(\Lambda_i^{-\frac{1}{2}}U_i^T\,{}^WR_{L_k}\right)^T\right]^{-1} \Lambda_i^{-\frac{1}{2}}U_i^T e_{ij}$$

$$= \left(e_{ij}\right)^T U_i\Lambda_i^{-\frac{1}{2}}\left[\Lambda_i^{\frac{1}{2}}U_i^T\,{}^WR_{L_k}\Sigma_n^{-1}\left({}^WR_{L_k}\right)^T U_i\Lambda_i^{\frac{1}{2}}\right]\Lambda_i^{-\frac{1}{2}}U_i^T e_{ij}$$

$$= \left(e_{ij}\right)^T \left[{}^WR_{L_k}\Sigma_n^{-1}\left({}^WR_{L_k}\right)^T\right]e_{ij}$$

$$= \frac{1}{\sigma}\left(e_{ij}\right)^T e_{ij}$$

(8.5)

Comparing (8.5) with the case without noise, the standardization term $\Lambda_i^{-\frac{1}{2}}U_i^T$ in the residual h_{ij} is offset by that in L_{ij} due to the introduction of noise, which is referred as constraint degeneration. Actually, the covariance component $\Lambda_i^{-\frac{1}{2}}U_i^T$ should be considered during optimization, due to the fact that the constraints provided by the local structures are uneven in the three principal component directions. Take a point-to-plane distribution correspondence as an example. For a point $\mathbf{p}$ of current frame, it is almost impossible that the point $\mathbf{p}$ coincides with the mean μ of corresponding plane distribution. Ideally, $\mathbf{p}$ and μ lie on the same plane, but there inevitably exist translation errors in the non-normal directions. This means that the confidence of the components of distance from point to distribution mean in the non-normal directions is lower than that in the normal direction. Therefore, we should pay different attentions on the three components of point-to-mean distance, and the component in the normal direction is endowed with the key role during optimization. The aforementioned idea can be satisfied by the covariance component $\Lambda_i^{-\frac{1}{2}} U_i^T$. Due to the disappearance of $\Lambda_i^{-\frac{1}{2}} U_i^T$ in degeneration results, the three distance components are treated equally and this reduces the localization accuracy.

For the point-to-distribution distance constraint term $\|h_{ij}\|_{\Sigma_o^{-1}}^2 = \left\|\Lambda_i^{-\frac{1}{2}} U_i^T e_{ij}\right\|_{\Sigma_o^{-1}}^2$ where Σ_o

is the covariance that describes the uncertainty of observation, and $\Lambda_i^{-\frac{1}{2}}$ plays the role of weighting the projected components $U_i^T e_{ij}$ of the error vector e_{ij} in the three principal directions. It acts like the loss function to reduce the impact of outliers in robust estimation. From this point of view, a special loss function $\rho\left(s, \Lambda_i^{-\frac{1}{2}}\right) = \left\|\Lambda_i^{-\frac{1}{2}} U_i^T e_{ij}\right\|_{\Sigma_o^{-1}}^2$ for point-to-distribution

distance is designed to balance the different components of the distance constraint with observation uncertainty, where $s = \|U_i^T e_{ij}\|_{\Sigma_o^{-1}}^2$. Different from the traditional loss functions, ρ is related to both the loss $\|U_i^T e_{ij}\|_{\Sigma_o^{-1}}^2$ as well as covariance component $\Lambda_i^{-\frac{1}{2}}$, and $\Lambda_i^{-\frac{1}{2}}$ is injected in $\|U_i^T e_{ij}\|_{\Sigma_o^{-1}}^2$ for considering different components. On this basis, the observation equation turns into $z_{ij} = U_i^T e_{ij} + v_i$, where $v_i = U_i^{T\,W} R_{L_k}{}^I R_L n \sim \left(0, U_i^{T\,W} R_{L_k} \Sigma_n \left(U_i^{T\,W} R_{L_k}\right)^T\right)$ and $\Sigma_o = U_i^{T\,W} R_{L_k} \Sigma_n \left(U_i^{T\,W} R_{L_k}\right)^T$. The point-to-distribution distance constraint is then converted to

$$\rho\left(\|U_i^T e_{ij}\|_{\Sigma_o^{-1}}^2, \Lambda_i^{-\frac{1}{2}}\right) = \left\|\Lambda_i^{-\frac{1}{2}} U_i^T e_{ij}\right\|_{\Sigma_o^{-1}}^2$$

$$= \left\|\Lambda_i^{-\frac{1}{2}} U_i^T e_{ij}\right\|_{\left(U_i^{T\,W} R_{L_k} \Sigma_n \left(U_i^{T\,W} R_{L_k}\right)^T\right)^{-1}}^2$$

$$= \left(e_{ij}\right)^T U_i \Lambda_i^{-\frac{1}{2}} \left[U_i^{T\,W} R_{L_k} \frac{1}{\sigma} I_{3\times3} \left({}^W R_{L_k}\right)^T U_i\right] \Lambda_i^{-\frac{1}{2}} U_i^T e_{ij} \qquad (8.6)$$

$$= \frac{1}{\sigma}\left(e_{ij}\right)^T U_i \Lambda_i^{-\frac{1}{2}} \Lambda_i^{-\frac{1}{2}} U_i^T e_{ij}$$

$$= \frac{1}{\sigma} e_{ij}^T \left({}^W \Sigma_i^C\right)^{-1} e_{ij}$$

$$= \|h_{ij}\|_{\Sigma_n^{-1}}^2$$

Still, the changed constraint contains the covariance component $\Lambda_i^{-\frac{1}{2}} U_i^T$ in h_{ij}, which means that the non-degenerate constraint is obtained.

Combining the prior state constraint (8.3) of IMU with (8.6), the cost function with respect to $\tilde{X}_k$ is built as follows:

$$\min_{\tilde{X}_{k,l}} \left\| \breve{X}_{k,l} \boxplus \tilde{X}_{k,l} \boxminus \bar{X}_k \right\|_{\bar{P}_k^{-1}}^2 + \sum_{i=1}^{N} \sum_{j=1}^{M_i} \sigma_i \left\| h_{ij} \left(\breve{X}_{k,l} \boxplus \tilde{X}_{k,l} \right) \right\|_{\Sigma_n^{-1}}^2 \tag{8.7}$$

where l and M_i represents the number of iterations and points in $\mathcal{S}_i$, respectively. $\sigma_i = \dfrac{\lambda_{i1} + \lambda_{i2}}{\lambda_{i1} + \lambda_{i2} + \lambda_{i3}}$ is the weight coefficient, where $\lambda_{i1}, \lambda_{i2}, \lambda_{i3}$ are the eigenvalues of $^W\Sigma_i^C$ and $\lambda_{i1} \leq \lambda_{i2} \leq \lambda_{i3}$. The cost function (8.7) can be solved with IEKF update as follows:

$$\breve{X}_{k,l+1} = \breve{X}_{k,l} \boxplus \tilde{X}_{k,l} \tag{8.8}$$

$$\tilde{X}_{k,l} = -S_{k,l} H_{k,l}^T R^{-1} h_{k,l} \left(\breve{X}_{k,l} \right) - \left(I - S_{k,l} H_{k,l}^T R^{-1} H_{k,l} \right) J_{k,l}^{-1} \left(\breve{X}_{k,l} \boxminus \bar{X}_k \right) \tag{8.9}$$

$$J_{k,l} = \begin{bmatrix} I_{6\times6} & 0_{6\times3} & 0_{6\times6} \\ 0_{3\times6} & J_r^{-1} \left({}^W\bar{R}_{I_k,l} \boxminus {}^W\bar{R}_{I_k,l} \right) & 0_{3\times6} \\ 0_{6\times6} & 0_{6\times3} & I_{6\times6} \end{bmatrix} \tag{8.10}$$

$$H_{k,l}^{ij} = \Lambda_i^{-\frac{1}{2}} U_i^T \begin{bmatrix} I_{3\times3} & 0_{3\times3} & -{}^W R_{I_k} \left[{}^{I_k} \mathbf{p}_{ij}^{\mathcal{S}} \right]_{\times} & 0_{3\times6} \end{bmatrix} \tag{8.11}$$

where ${}^W\bar{R}_{I_k,l}$ refers to the rotation component in the predicted state $\bar{X}_k$. $H_{k,l} = \begin{bmatrix} \cdots & \left(\sqrt{\sigma_i} H_{k,l}^{ij} \right)^T & \cdots \end{bmatrix}^T$ and $h_{k,l} \left(\breve{X}_{k,l} \right) = \begin{bmatrix} \cdots & \left(\sqrt{\sigma_i} h_{ij} \left(\breve{X}_{k,l} \right) \right)^T & \cdots \end{bmatrix}^T$. $J_{k,l}$ and $H_{k,l}^{ij}$ are the Jacobians of $\breve{X}_{k,l} \boxplus \tilde{X}_{k,l} \boxminus \tilde{X}_{k,l}$ and $h_{ij} \left(\breve{X}_{k,l} \boxplus \tilde{X}_{k,l} \right)$ with respect to $\tilde{X}_{k,l}$ evaluated at zero, respectively. $J_r(\cdot)$ is the right Jacobian of $SO(3)$ [22]. $S_{k,l} = \left(J_{k,l}^{-T} \bar{P}_k^{-1} J_{k,l}^{-1} + H_{k,l}^T R^{-1} H_{k,l} \right)^{-1}$ and R is the block diagonal matrix constituting of N noise covariance Σ_n. Equations (8.10) and (8.11) are solved with the decoupled strategy in the Chapter 7.

At the beginning of the iteration, $\breve{X}_{k,0}$ is set to the prior state $\bar{X}_k$. During iteration, the data association is re-performed at a fixed interval N_I for more accurate correspondences.

When the state estimation converges or the maximum iteration $l_{\max}$ is reached, the optimal state estimation $\hat{X}_k$ is obtained with the covariance updated as below:

$$\hat{P}_k = \left(I - S_k H_k^T R^{-1} H_k \right) \bar{P}_k \tag{8.12}$$

where S_k and H_k adopt the values at the last iteration.

In this low-level LIO, a local map needs to be maintained for data association of distributions. When the distance between current frame and the frame that is added to the local map in the last time exceeds the preset threshold d_{th} or the ratio of number of correspondences to that of source distributions is smaller than r_{th}, the point cloud of current frame is registered for local map maintenance.

Algorithm 8.1 describes the entire process of the low-level LIO. For the first $\mathcal{F}_{\mathrm{th}}$ frames, initialization [23] based on IMU pre-integration [5] is executed to get the velocities, biases of IMU, and gravity in the world frame. During initialization, the relative poses of two adjacent frames are estimated by the LiDAR scan-to-local map. And the estimated results serve as the outputs of the LIO. $^{W}p_{\mathrm{pre}}$ is the position of the last frame added into local map. ε is the threshold to judge iteration convergence. $RadiusFilterAndTransform\,(\cdot)$ is used to filter invalid points around the LiDAR center and then project the point cloud $^{L_k}\mathcal{P}$ from LiDAR frame to IMU frame. $Predict\,(\cdot)$ and $CorrectPointCloud\,(\cdot)$ perform the prior pose prediction and motion distortion correction with IMU measurements, respectively. $ExtractDistribution\,(\cdot)$ executes distribution extraction and $DataAssociation\,(\cdot)$ represents data association between the source distributions and the local map.

Algorithm 8.1 Low-Level LiDAR Inertial Odometry

Input: point cloud $^{L_k}\mathcal{P}$, IMU measurements $\left\{ a_q, \omega_q \mid 1 \le q \le \mathbf{q} \right\}_{(t_{k-1}, t_k]}$, local map $\mathcal{M}$, $\hat{X}_{k-1}$, and $\hat{P}_{k-1}$;

Output: state estimation $\hat{X}_k$ and covariance $\hat{P}_k$.

1 $^{I_k}\mathcal{P} = RadiusFilterAndTransform\left(^{L_k}\mathcal{P} \right)$;
2 downsample $^{I_k}\mathcal{P}$;
3 **if** $k \le \mathcal{F}_{\mathrm{th}}$ **do**
4   **if** $k == 1$ **do**
5       $\hat{X}_k = \begin{bmatrix} 0_{3\times2} & I_{3\times3} & 0_{3\times2} \end{bmatrix}^T$;
6   initialize $\mathcal{M}$ with $\hat{X}_k$;
7   **return**
8   **end if**
9   calculate $\hat{X}_k$ with $^{I_k}\mathcal{P}$ and $\mathcal{M}$, and update local map $\mathcal{M}$;
10   calculate IMU pre-integration γ_k;
11   **if** $k == \mathcal{F}_{\mathrm{th}}$ **do**
12   initialize IMU with $\{\hat{X}_k \mid 1 \le k \le \mathcal{F}_{\mathrm{th}}\}$ and $\{\gamma_k \mid 1 \le k \le \mathcal{F}_{\mathrm{th}}\}$;
13   **end if**

14 **else do**
15 $\left\{\left\{\bar{X}_\tau \mid \tau \in (t_{k-1}, t_k]\right\}, \bar{X}_k, \bar{P}_k\right\} = Predict\left(\left\{a_q, \omega_q \mid 1 \le q \le \mathbf{q}\right\}_{(t_{k-1}, t_k]}, \hat{X}_{k-1}, \hat{P}_{k-1}\right);$
16 $^{I_k}\mathcal{P} = CorrectPointCloud\left(^{I_k}\mathcal{P}, \left\{\bar{X}_\tau \mid \tau \in (t_{k-1}, t_k]\right\}\right);$
17 $\left\{\mathcal{S}_{i_s} \mid 1 \le i_s \le N_s\right\} = ExtractDistribution\left(^{I_k}\mathcal{P}\right);$
18 $X_{k,0} = \bar{X}_k;$
19 **for** $l \in \{0, 1, \ldots, l_{\max} - 1\}$ **do**
20   **if** $l \% N_I == 0$ **do**
21      $\left\{(\mathcal{S}_i, \mathcal{T}_i) \mid 1 \le i \le N\right\} = DataAssociation\left(^{I_k}\mathcal{P}, \breve{X}_{k,l}, \mathcal{M}\right);$
22   **end if**
23   calculate $J_{k,l}^{-1}$ and $H_{k,l}$ with $\left\{(\mathcal{S}_i, \mathcal{T}_i)\right\}$ and $\breve{X}_{k,l};$
24   calculate $\tilde{X}_{k,l}$ and $\breve{X}_{k,l+1};$
25   **if** $\left\|\tilde{X}_{k,l} - \tilde{X}_{k,l-1}\right\| < \varepsilon$ or $l == l_{\max} - 1$ **do**
26      $\hat{X}_k = \breve{X}_{k,l+1};$
27   **break;**
28   **end if**
29 **end for**
30 calculate $\hat{P}_k;$
31 **if** $\left\|^W p_k - {}^W p_{\mathrm{pre}}\right\| \ge d_{\mathrm{th}}$ or $N/N_s < r_{\mathrm{th}}$ **do**
32   update local map $\mathcal{M};$
33 **end if**
34 **end if**
35 **return**

8.3.3 High-Level Factor Graph Optimization

To further improve the local consistence of trajectory, we introduce a factor graph optimization after the LIO. It constructs cross-constraints among multiple frames with the direct measurements from both LiDAR and IMU. The poses within the sliding window are simultaneously optimized, which spreads the accumulated error to the local trajectory and improves the consistence and precision.

8.3.3.1 Prior Factor

To decrease the computational burden, the sliding window strategy is adopted. Only the recently selected L frames are maintained in the graph. Once the number of frames within the window exceeds L, the oldest frame is removed. Its associated constraints are marginalized via Schur-complement [24] to constitute a prior factor for avoiding the information loss. This factor is activated when there are L frames within the sliding window.

8.3.3.2 IMU Pre-Integration Factor

The pre-integration factor is composed of the pre-integration result [5] of IMU measurements between two frames, which provides a relative motion constraint between the corresponding poses. Without repeated integration, the pre-integration supports an effective joint optimization for IMU measurements with other sensors. Note that the initial values of IMU parameters for factor graph optimization are the same as those in IMU initialization of low-level LIO.

8.3.3.3 LiDAR Observation Constraint

For the LiDAR point cloud, we directly construct point-to-distribution distance constraints on the points from different frames instead of the relative poses from the low level. To construct constraints, the corrected point clouds are projected to the local voxel map with the estimation from the low-level EKF. The points falling into the same voxel from different frames are assumed to follow the Gaussian distribution described by the union of these points. Their log-likelihoods are incorporated to form the cross-constraints among different frames. Also, the weight coefficient corresponding to the LiDAR observation constraints in a voxel is calculated according to the covariance of all points. When the coefficient is less than a certain threshold, the voxel is discarded without participating in optimization. Otherwise, the constraints are weighted by the coefficient. Moreover, another anti-outlier weight w_l related to residual is attached to reduce the influence of outliers. $w_l = \dfrac{\rho_o}{\rho_o + C}$, where C is average cost of all points from one frame in a voxel and ρ_o is a preset constant. The larger the average cost C is, the more likely outliers are. With the increasing of C, the weight w_l decreases, which limits the interference caused by outliers. When the window slides, the point cloud of the oldest frame in the voxel map is marginalized and the corresponding state is deleted.

By integrating the estimated results from low and high levels, the final odometry output is obtained. These two estimations are executed in two independent threads and the output frequency of the proposed hierarchical framework is consistent with that of low level.

8.4 EXPERIMENTS

To demonstrate the effectiveness of the proposed hierarchical distribution-based LIO method termed as HD-LIO, we conduct the validation on the public Newer College (NC) dataset [25] and the extended dataset (ENC) [26]. The NC dataset collects point clouds using the handheld Ouster OS1–64 from structured buildings, vegetation, and open space areas with lack of texture. It contains five sequences with fast walking and aggressive motions of the device. In the ENC dataset, the laser scanner is replaced with Ouster OS0–128. Moreover, compared to NC dataset, ENC covers new scenes and different levels of difficulty are especially involved according to the aggressiveness of the motion. For both two datasets, the built-in IMU of Ouster is employed to provide inertial measurements. All the following experiments are conducted on a PC with an Intel Core i7–9750H CPU.

8.4.1 Ablation Experiment

To verify the performance of the proposed HD-LIO method, its three variants are involved according to whether anti-degeneration constraint, factor graph optimization as well as anti-outlier weight w_l are considered. HD-LIO-I and HD-LIO-II refer to the low-level LIO without or with anti-degeneration constraints, respectively, while HD-LIO-III corresponds to the HD-LIO without the anti-outlier weight. Table 8.1 presents the accuracy results of different

TABLE 8.1   Ablation Study of Different Variants of the Proposed HD-LIO Method on the ENC and NC Datasets

		HD-LIO-I		HD-LIO-II		HD-LIO-III		HD-LIO	
Dataset	**Sequence**	RTE	RRE	RTE	RRE	RTE	RRE	RTE	RRE
ENC	Cloister	0.79	1.61	0.34	1.01	0.36	1.09	0.28	0.91
	Park	0.54	0.40	0.30	0.28	0.33	0.29	0.31	0.29
	Quad-easy	0.30	0.84	0.23	0.87	0.21	0.86	0.21	0.87
	Quad-medium	0.44	1.17	0.34	1.37	0.33	1.34	0.33	1.35
	Quad-hard	1.81	3.25	0.58	1.82	0.57	1.65	0.58	1.46
	Maths-easy	0.61	1.07	0.50	1.09	0.46	0.96	0.46	0.96
	Maths-medium	0.87	1.81	0.82	1.79	0.82	1.78	0.82	1.78
	Maths-hard	0.40	1.08	0.31	0.83	0.29	0.80	0.30	0.81
NC	Short experiment	1.23	1.30	0.94	1.10	0.93	1.09	0.92	1.10
	Long experiment	1.12	1.37	0.91	1.22	0.92	1.22	0.92	1.22
	Quad with dynamics	0.25	1.35	0.22	1.33	0.22	1.32	0.22	1.31
	Dynamic spinning	0.59	12.08	0.53	12.19	0.53	12.16	0.52	12.17
	Parkland mound	0.89	1.90	0.55	1.43	0.50	1.36	0.49	1.33
Average		0.76	2.25	0.51	2.03	0.50	1.99	0.49	1.97

RRE, relative rotation error (°/100 m); RTE, relative translation error (%).

variants on the ENC and NC datasets in terms of RTE and RRE. It is observed from the results of HD-LIO-I and HD-LIO-II that both translation and rotation errors are significantly reduced, which shows the advantage of the anti-degeneration solution. Also, the introduction of the high-level factor graph optimization with anti-outlier weight is helpful for localization.

8.4.2 Accuracy Comparison with Existing Methods

In this section, we conduct a comparison with existing LIO/SLAM methods on the ENC and NC datasets. The comparative methods are summarized as follows.

- **LINS** [12]: The tight-coupled LIO that solves the 6-DOF ego-motion via filtering;

- **CLINS** [27]: A continue-time LiDAR inertial SLAM combining a non-rigid registration and a global pose graph optimization with loop closure;

- **LiLi-OM** [19]: A high-performance solid-state LiDAR inertial method with a lightweight LiDAR odometry frontend and a keyframe-based graph optimization including loop closure;

- **FAST-LIO2** [13]: An IEKF-based LIO framework relying on the scan-to-global map matching;

- LINS w/o loop closure, CLINS w/o loop closure, and LiLi-OM w/o loop closure: The variants of corresponding methods with the loop closure deactivated.

The comparison results are listed in Table 8.2. Comparing the results of the proposed HD-LIO method with those of LINS, CLINS, LiLi-OM as well as their versions without

TABLE 8.2 Accuracy Comparison of Different Methods on the ENC and NC Datasets in Terms of RTE (%) and RRE (°/100 m)

Dataset	Sequence	LINS w/o Loop Closure		LINS		CLINS w/o Loop Closure		CLINS		LiLi-OM w/o Loop Closure		LiLi-OM		FAST-LIO2		HD-LIO	
		RTE	RRE	RTE	RRE	RTE	RRE	RTE	RRE	RTE	RRE	RTE	RRE	RTE	RRE	RTE	RRE
ENC	Cloister (429 m)	0.86	2.12	0.77	1.99	0.66	1.61	0.53	1.39	×	×	×	×	0.17	0.52	0.28	0.91
	Park (2,396 m)	0.92	1.14	0.90	1.11	2.93	2.46	2.86	2.76	×	×	×	×	0.24	0.24	0.31	0.29
	Quad-easy (247 m)	0.56	1.89	0.59	1.92	0.39	1.16	0.49	1.81	0.81	1.55	0.58	1.53	0.23	0.89	0.21	0.87
	Quad-medium (260 m)	0.75	2.70	0.76	2.66	0.61	1.96	0.43	1.96	0.57	2.12	0.59	2.13	0.35	1.38	0.33	1.35
	Quad-hard (234 m)	×	×	×	×	0.76	1.90	0.74	2.14	1.18	2.49	0.90	2.38	0.61	1.52	0.58	1.46
	Maths-easy (264 m)	1.04	2.40	1.06	2.40	0.76	1.53	0.72	1.64	0.98	1.75	0.98	1.74	0.41	0.95	0.46	0.96
	Maths-medium (304 m)	1.49	3.44	1.52	3.33	0.77	1.90	0.77	1.89	1.02	2.00	1.02	2.00	0.76	1.68	0.82	1.78
	Maths-hard (321 m)	4.82	17.38	4.48	15.34	0.62	1.51	0.69	1.74	1.29	2.41	1.29	2.40	0.34	0.83	0.30	0.81
NC	Short experiment (1,609 m)	1.06	1.24	1.08	1.26	6.53	5.73	9.73	8.39	0.97	1.05	0.94	1.04	0.93	1.06	0.92	1.10
	Long experiment (3,063 m)	1.07	1.46	1.09	1.48	4.86	4.38	2.06	2.54	0.92	1.22	0.90	1.19	0.89	1.18	0.92	1.22
	Quad with dynamics (479 m)	×	×	×	×	×	×	×	×	0.31	1.56	0.31	1.54	0.24	1.31	0.22	1.31
	Dynamic spinning (97 m)	×	×	×	×	0.67	12.76	0.79	13.80	0.62	18.12	0.61	17.02	0.51	12.18	0.52	12.17
	Parkland mound (696 m)	0.71	1.71	0.73	1.75	×	×	×	×	0.79	1.52	0.59	1.47	0.50	1.36	0.49	1.33

×, localization failure.

loop closure, HD-LIO performs well. Specifically, our method gains advantage in the sequences of quad-easy, quad-medium, parkland mound, quad-hard, maths-hard, and quad with dynamics. It is worth noticing that the latter three sequences as well as dynamic spinning are challenging as there exist aggressive motions. Figure 8.2 shows the angular velocity and acceleration curves of device for the dynamic spinning and maths-hard sequences. The maximum rotation rate and acceleration even reach up to 249.9°/s and 14.5 m/s², respectively. The overall result under these extreme motion conditions indicates the effectiveness of our method.

FAST-LIO2 demonstrates the excellent performance and it performs better than our HD-LIO in the sequences of cloister, maths-easy, and maths-medium. The structured scenarios are beneficial for point-to-plane constraints of FAST-LIO2. In particular, it manages a global map for matching, which implies implicit loop closure. This promotes its performance on the sequences of park and long experiments with many

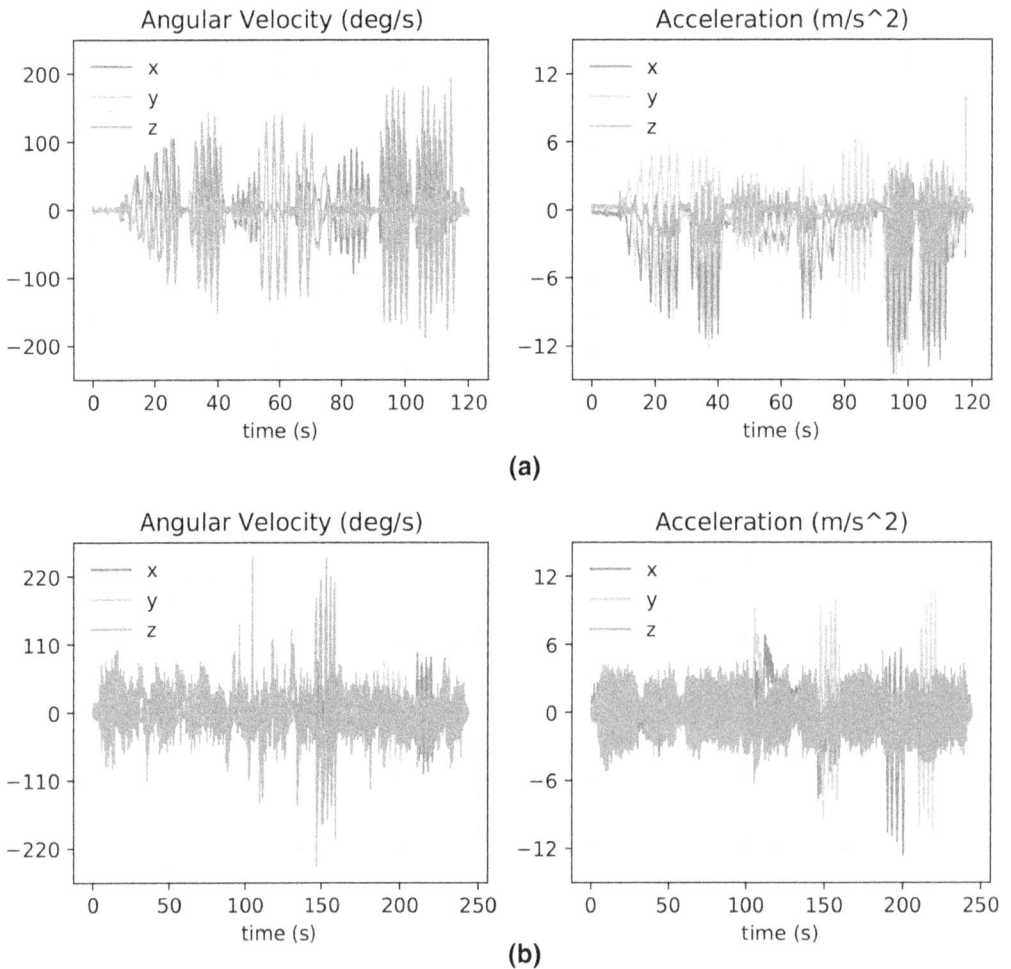

FIGURE 8.2 The angular velocity and acceleration curves of device. (a) Dynamic spinning sequence in the NC dataset. (b) Maths-hard sequence in the ENC dataset.

places re-visited. By contrast, our HD-LIO method is only based on local map without any loop closure, which benefits less from these scenarios. To further demonstrate the proposed method, we split the park sequence into five independent segments, which correspond to the time intervals [0, 385), [385, 730), [730, 1,110), [1,110, 1,300), and [1,300, 1,572] in seconds. Segments 1–3 contain loops and there is no loop for segments 4 and 5. On each segment, FAST-LIO2 and our HD-LIO are individually run and the results are presented in Table 8.3. The trajectories of these two methods on each segment of the park sequence are depicted in Figure 8.3. It is indicated that FAST-LIO2

TABLE 8.3 Accuracy Comparison of FAST-LIO2 and HD-LIO on the Five Independent Segments of the Park Sequence

Sequence	Segment	FAST-LIO2		HD-LIO	
		RTE	RRE	RTE	RRE
Park	0–385 s	0.19	0.36	0.18	0.36
	385–730 s	0.50	0.49	0.50	0.46
	730–1,110 s	0.39	0.35	0.54	0.48
	1,110–1,300 s	2.98	2.74	0.66	0.59
	1,300–1,572 s	2.08	1.34	0.78	0.68

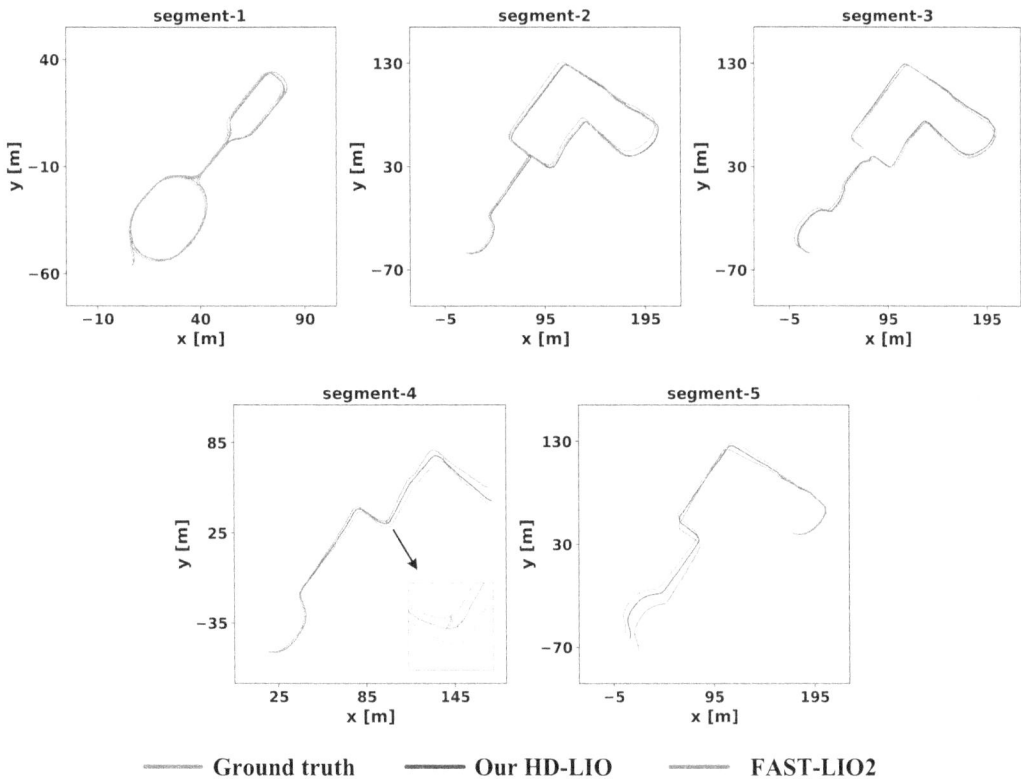

FIGURE 8.3 The trajectories of HD-LIO and FAST-LIO2 on the five independent segments of the park sequence in the ENC dataset.

performs well in the former three segments. For the latter two segments, there are no loops and HD-LIO performs better. Actually, there exist weakly structured positions in segments 4 and 5. In these positions, one side of the road is full of geometric structures, whereas the other side is covered with vegetation. In this case, FAST-LIO2 which mainly depends on the plane features, drifts slightly. By contrast, our method employs distribution and keeps it stable. This is the reason that HD-LIO has better-estimated trajectories in these two segments.

Figure 8.4 illustrates the trajectories of different methods on a sequence of NC dataset. It can be observed that the trajectory estimated by our HD-LIO is close to the ground truth. Figure 8.5 provides the mapping results for our method, CLINS, LiLi-OM, and FAST-LIO2 on the maths-hard sequence of the ENC dataset, where the variant of CLINS without loop closure and the degeneration variant HD-LIO-I of our method are also involved. One can see that there is visible ghosting (see local enlarged parts) in the point cloud maps of CLINS variant, as illustrated in the first row of Figure 8.5. With the introduction of loop closure, the ghosting problem is solved for CLINS with a consistent mapping result. LiLi-OM still faces slight inconsistency despite the support of loop closure. For the variant HD-LIO-I of our method, the ghosting also exists due to the degeneration problem. After anti-degeneration solution is introduced, the proposed HD-LIO (only using local map without any loop closure) achieves a consistent mapping result, which is the same as that of FAST-LIO2 that relies on the global map.

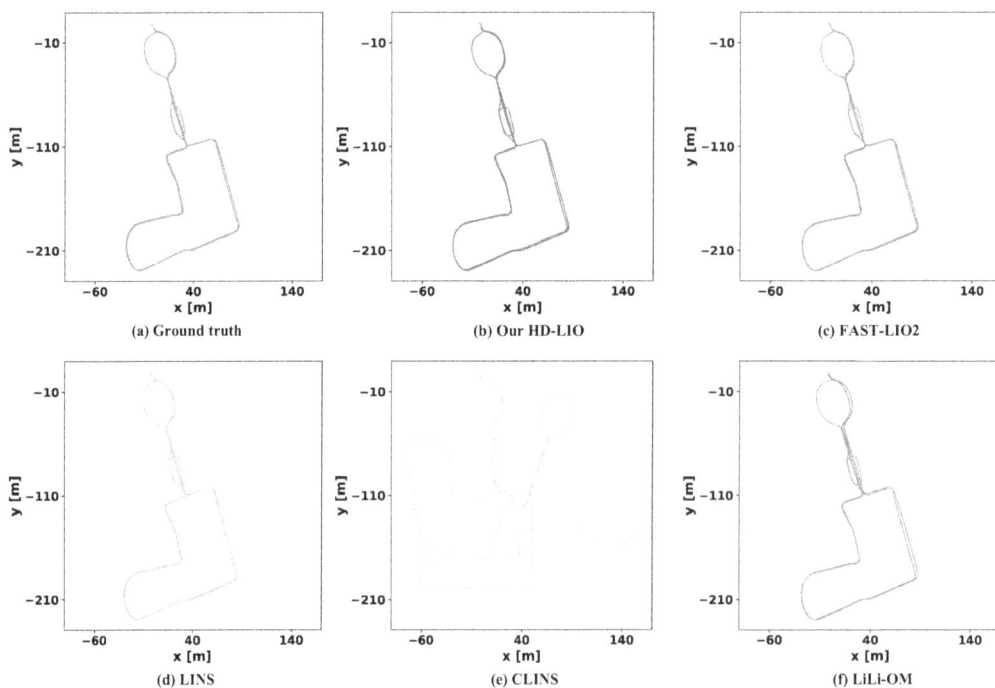

FIGURE 8.4   The trajectories of different methods on a sequence of NC dataset.

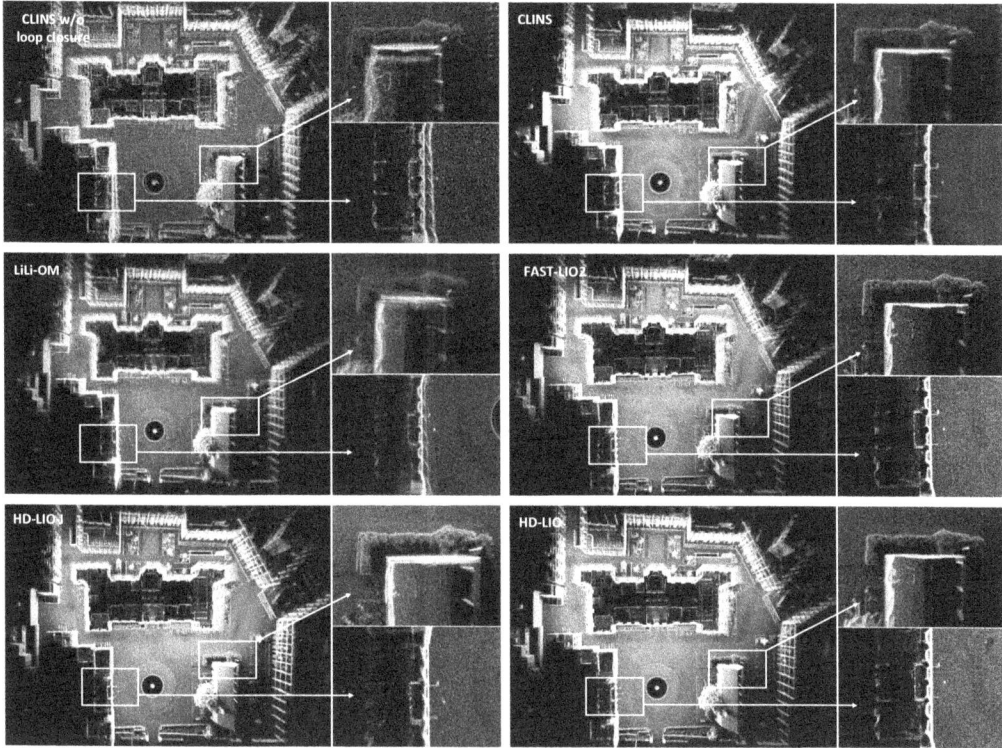

FIGURE 8.5   The mapping results of different methods on the maths-hard sequence of the ENC dataset.

TABLE 8.4   Processing Times (ms) of the Proposed Method on ENC and NC Datasets

	NC		ENC	
	Avg	Std	Avg	Std
Low-level LiDAR inertial odometry	44.39	10.18	45.85	13.28
High-level factor graph optimization	104.65	21.03	133.48	17.35

Avg, average value; Std, standard deviation.

8.4.3 Efficiency Analysis

Next, the efficiency of the proposed method is analyzed. Our HD-LIO consists of a low-level LIO and a high-level factor graph optimization. They run in two separate threads and the low-level thread decides the output frequency of the whole method. The processing time on the NC and ENC datasets is given in Table 8.4. As we can see, the average processing time of the low-level thread is less than 100 ms (the scanning period of LiDAR), which shows the real-time characteristic of the proposed HD-LIO. Besides, for the high-level thread, it is run every four frames and its processing time less than 400 ms makes the graph optimization to be fully executed. Figure 8.6 shows the frame-wise processing time of the low-level and high-level

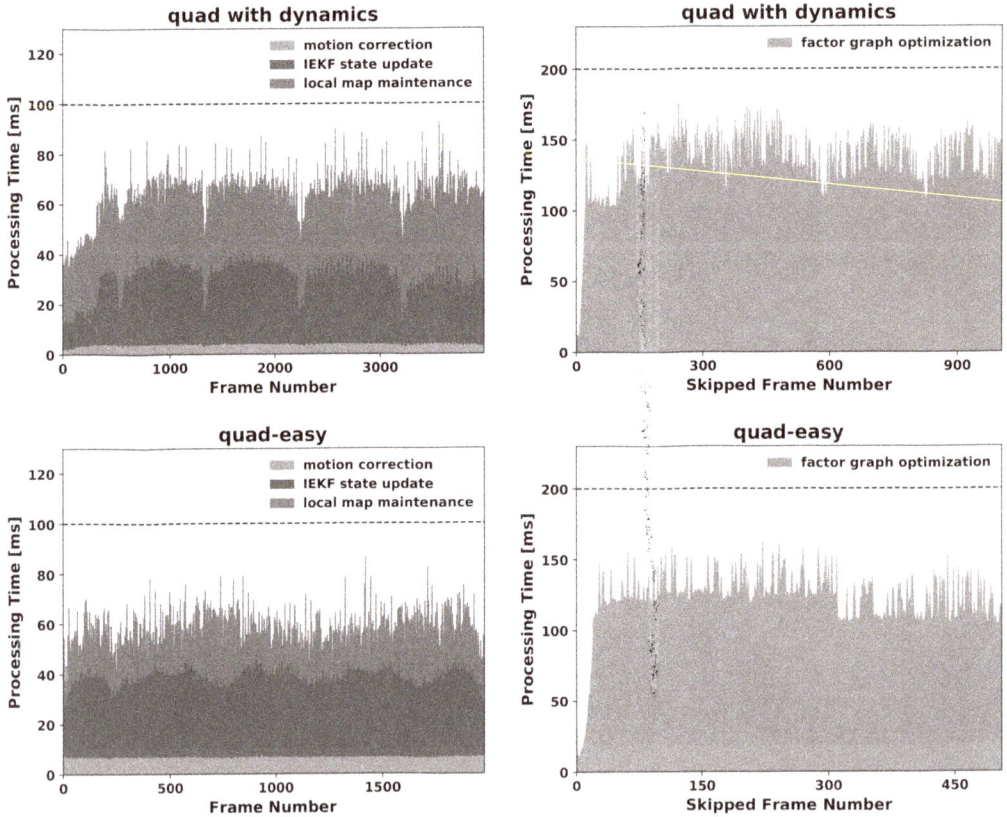

FIGURE 8.6 The frame-wise processing time of the proposed method on the two selected sequences of NC and ENC datasets.

threads on the selected sequences of NC and ENC datasets, where the former includes three stages: motion correction, IEKF state update, and local map maintenance. One can see from the result of low-level thread that each frame of the proposed method can be processed in real time, no matter 64 channels (NC) or 128 channels (ENC) of LiDAR resolution.

8.5 CONCLUSION

This chapter presents a hierarchical tightly coupled LIO based on distribution, which is composed of a low-level IEKF-based LIO and a high-level factor graph optimization. Aiming at the point cloud constraint degeneration problem in the distribution-based LIO due to the uncertainty of LiDAR measurement noise, an anti-degeneration solution is designed by the rectification of residual uncertainty, ensuring the effective tight coupling of point cloud and inertial measurements. Besides, the factor graph optimization with a prior factor, IMU pre-integration factors, and LiDAR observation constraints among multiple frames further refine the odometry estimation. All these efforts yield real-time precise odometry results. The experiment results on the NC and ENC datasets demonstrate the advantage of the proposed method.

REFERENCES

[1] Bai, C., Xiao, T., Chen, Y., Wang, H., Zhang, F., & Gao, X. (2022). Faster-LIO: Lightweight tightly coupled Lidar-inertial odometry using parallel sparse incremental voxels. *IEEE Robotics and Automation Letters*, 7(2), 4861–4868.

[2] Ye, H., Chen, Y., & Liu, M. (2019). Tightly coupled 3D Lidar inertial odometry and mapping. In *Proceedings of the IEEE International Conference on Robotics and Automation*, Montreal, QC, Canada (pp. 3144–3150).

[3] Zhang, J., & Singh, S. (2014). LOAM: Lidar odometry and mapping in real-time. *Robotics: Science and Systems*, 2(9), 1–9.

[4] Chen, K., Lopez, B. T., Agha-mohammadi, A. A., & Mehta, A. (2022). Direct LiDAR odometry: Fast localization with dense point clouds. *IEEE Robotics and Automation Letters*, 7(2), 2000–2007.

[5] Forster, C., Carlone, L., Dellaert, F., & Scaramuzza, D. (2017). On-manifold preintegration for real-time visual-inertial odometry. *IEEE Transactions on Robotics*, 33(1), 1–21.

[6] Palieri, M., Morrell, B., Thakur, A., Ebadi, K., Nash, J., Chatterjee, A., Kanellakis, C., Carlone, L., Guaragnella, C., & Agha-Mohammadi, A. A. (2021). LOCUS: A multi-sensor lidar-centric solution for high-precision odometry and 3D mapping in real-time. *IEEE Robotics and Automation Letters*, 6(2), 421–428.

[7] Reinke, A., Palieri, M., Morrell, B., Chang, Y., Ebadi, K., Carlone, L., & Agha-Mohammadi, A. A. (2022). LOCUS 2.0: Robust and computationally efficient lidar odometry for real-time 3D mapping. *IEEE Robotics and Automation Letters*, 7(4), 9043–9050.

[8] Kubelka, V., Vaidis, M., & Pomerleau, F. (2022). Gravity-constrained point cloud registration. In *Proceedings of the IEEE/RSJ International Conference on Intelligent Robots and Systems*, Kyoto, Japan (pp. 4873–4879).

[9] Hess, W., Kohler, D., Rapp, H., & Andor, D. (2016). Real-time loop closure in 2D LIDAR SLAM. In *Proceedings of the IEEE International Conference on Robotics and Automation*, Stockholm, Sweden (pp. 1271–1278).

[10] Hening, S., Ippolito, C. A., Krishnakumar, K., Stepanyan, V., & Teodorescu, M. (2017). 3D LiDAR SLAM integration with GPS/INS for UAVs in urban GPS-degraded environments. In *AIAA Information Systems-AIAA Infotech@Aerospace*, doi: 10.2514/6.2017-0448.

[11] Jiao, J., Zhu, Y., Ye, H., Huang, H., Yun, P., Jiang, L., Wang, L., & Liu, M. (2021). Greedy-based feature selection for efficient LiDAR SLAM. In *Proceedings of the IEEE International Conference on Robotics and Automation*, Xi'an, China (pp. 5222–5228).

[12] Qin, C., Ye, H., Pranata, C., Han, J., Zhang, S., & Liu, M. (2020). LINS: A Lidar-inertial state estimator for robust and efficient navigation. In *Proceedings of the IEEE International Conference on Robotics and Automation*, Paris, France (pp. 8899–8906).

[13] Xu, W., Cai, Y., He, D., Lin, J., & Zhang, F. (2022). FAST-LIO2: Fast direct LiDAR-inertial odometry. *IEEE Transactions on Robotics*, 38(4), 2053–2073.

[14] Xu, W., & Zhang, F. (2021). FAST-LIO: A Fast, robust LiDAR-inertial odometry package by tightly-coupled iterated Kalman filter. *IEEE Robotics and Automation Letters*, 6(2), 3317–3324.

[15] Hesch, J., Mirzaei, F., Mariottini, G., & Roumeliotis, S. (2010). A Laser-Aided Inertial Navigation System (L-INS) for human localization in unknown indoor environments. In *Proceedings of the IEEE International Conference on Robotics and Automation*, Anchorage, AK, USA (pp. 5376–5382).

[16] Shan, T., Englot, B., Meyers, D., Wang, W., Ratti, C., & Rus, D. (2020). LIO-SAM: Tightly-coupled Lidar inertial odometry via smoothing and mapping. In *Proceedings of the IEEE/RSJ International Conference on Intelligent Robots and Systems*, Las Vegas, NV, USA (pp. 5135–5142).

[17] Chen, K., Nemiroff, R., & Lopez, B. (2023). Direct LiDAR-inertial odometry: Lightweight LIO with continuous-time motion correction. In *Proceedings of the IEEE International Conference on Robotics and Automation*, London, United Kingdom (pp. 3983–3989).

[18] Wang, Z., Zhang, L., Shen, Y., & Zhou, Y. (2023). D-LIOM: Tightly-coupled direct LiDAR-inertial odometry and mapping. *IEEE Transactions on Multimedia*, 25, 3905–3920.

[19] Li, K., Li, M., & Hanebeck, U. (2021). Towards high-performance solid-state-LiDAR-inertial odometry and mapping. *IEEE Robotics and Automation Letters*, 6(3), 5167–5174.

[20] Li, W., Hu, Y., Han, Y., & Li, X. (2021). KFS-LIO: Key-feature selection for lightweight Lidar inertial odometry. In *Proceedings of the IEEE International Conference on Robotics and Automation*, Xi'an, China (pp. 5042–5048).

[21] Solà, J. (2017). Quaternion kinematics for the error-state Kalman filter. arXiv preprint arXiv:1711.02508.

[22] Solà, J., Deray, J., & Atchuthan, D. (2021). A micro Lie theory for state estimation in robotics. arXiv preprint arXiv:1812.01537v9.

[23] Qin, T., Li, P., & Shen, S. (2018). VINS-Mono: A robust and versatile monocular visual-inertial state estimator. *IEEE Transactions on Robotics*, 34(4), 1004–1020.

[24] Wei, W., Zhu, X., & Wang, Y. (2022). Novel robust simultaneous localization and mapping for long-term autonomous robots. *Frontiers of Information Technology & Electronic Engineering*, 23(2), 234–245.

[25] Ramezani, M., Wang, Y., Camurri, M., Wisth, D., Mattamala, M., & Fallon, M. (2020). The Newer College Dataset: Handheld LiDAR, inertial and vision with ground truth. In *Proceedings of the IEEE/RSJ International Conference on Intelligent Robots and Systems*, Las Vegas, NV, USA (pp. 4353–4360).

[26] Zhang, L., Camurri, M., Wisth, D., & Fallon, M. (2022). Multi-camera LiDAR inertial extension to the newer college dataset. arXiv preprint arXiv:2112.08854v3.

[27] Lv, J., Hu, K., Xu, J., Liu, Y., Ma, X., & Zuo, X. (2021). CLINS: Continuous-time trajectory estimation for LiDAR-inertial system. In *Proceedings of the IEEE/RSJ International Conference on Intelligent Robots and Systems*, Prague, Czech Republic (pp. 6657–6663).

CHAPTER 9

LiDAR Place Recognition Based on Range Image and Column-Shift-Invariant Attention

9.1 INTRODUCTION

Given a query observation, LiDAR place recognition (LPR) refers to identifying the same place from a reference database that contains a series of historical observations. Existing methods can be generally classified into local-descriptor-based and global-descriptor-based types. The former extracts local features from LiDAR observations and determines the target place by feature matching [1–3]. However, it lacks a global perspective, affecting the recognition performance. In contrast, the global-descriptor-based solution converts the LiDAR observation to global description vector and the place recognition problem is thus regarded as the retrieval based on the similarity of global descriptors [4–7]. It becomes mainstream due to high precision and efficiency through the exploitation of global information and retrieval technology. With the excellent representation ability of deep networks, the learning-based methods [6,7] generally perform better than those based on statistics [4,5].

A robust learning-based LPR method requires the recognition result to be invariant to the yaw rotation of query point cloud. This means that the retrieved results of the query point cloud and its rotated ones around the yaw-axis should be consistent. Taking original point cloud as input, Refs. [6,8,9] depend on data augmentation or extra transformation networks to realize the rotation invariance. Nevertheless, this solution is time-consuming with weak robustness. Recently, Ma et al. proposed a yaw-angle-invariant transformer network OverlapTransformer, where a projected range image from a point cloud is taken as input [7]. It utilizes an equivariant feature encoder and a self-attention module [10] to

DOI: 10.1201/9781003643630-9

extract feature maps, which shifts equally with the range image, and then an invariant NetVLAD pooling [11] is adopted to compress the features into a global description vector. With the invariance of the network to the horizontal shift of range images, it achieves good robustness to rotation. On the premise of depending on the invariance of NetVLAD, a possible problem is its equivariant feature extraction that is implemented by fixing the width size of feature maps. This restricts the change of feature scale before NetVLAD and thus affects the recognition accuracy. Actually, the information along the horizontal direction of the point cloud (i.e., width of the range image) is significant, which can be effectively leveraged by mining multi-scale features with the robustness to horizontal shift of range image. It is worth mentioning that the existing multi-scale features in vision often adopt strided convolution or pooling to reduce dimension and mainly mix multi-scale information along channel, which is not suitable for input features without spatial continuity.

To fully mine features, a LPR method based on range image and column-shift-invariant (CSI) attention is presented. Instead of relying on NetVLAD for invariance, we adopt CSI attention to realize invariance before NetVLAD. This early implementation of invariance enables the variation of feature scales before NetVLAD, and thus, the application of multi-scale features with dimension reduction becomes a reality. Further, an enhanced global descriptor is obtained with the robustness of the rotation of the point cloud.

The main contributions of this chapter are as follows. A global descriptor extraction network based on range image is proposed for LPR, where the CSI attention and a multi-scale feature module are cascaded to obtain global descriptors with good representation ability and robustness. As a result, the accuracy of place recognition is guaranteed with the adaptability to viewpoint changes of the point cloud. By applying circular convolution on the pooled input features, the CSI attention prompts the attention map to shift synchronously with the input features, which enables the attention to achieve the invariance under the weighted form of matrix multiplication. This invariant design makes the subsequent network be free from the scale restrictions, which is beneficial to exploiting features at different scales while capturing the global contextual information. Besides, a multi-scale feature module is used to mine the multi-scale information from the spatially discontinuous outputs of CSI attention, where multiple fully connected (FC) layers and a residual block with channel/width mixing are leveraged for feature downsampling and multi-scale information mixing, respectively. The incorporation of more information enriches the representation of features and enhances the discriminability of the global descriptors, which facilitates the performance of place recognition. The experiments on the KITTI, Ford Campus, and NCLT datasets demonstrate the effectiveness of the proposed method.

The rest of the chapter is organized as follows. Section 9.2 presents the related work. Section 9.3 describes the proposed method in detail. The experimental verification is shown in Section 9.4, and Section 9.5 concludes this chapter.

9.2 REVIEW OF LIDAR PLACE RECOGNITION

This section discusses the LPR from two aspects: local-descriptor-based and global-descriptor-based methods.

9.2.1 Local-Descriptor-Based Methods

Commonly, this type explicitly extracts local features from specific regions of a LiDAR scan and then suggests placing candidates according to feature matching. Guo et al. [1] extracted ISHOT (intensity signature of histograms of orientations) descriptors around the neighborhoods of the detected keypoints, which are combined with the voting model to determine possible matches of a query place. Steder et al. [2] employed bag-of-words and feature-matching evaluation based on normal aligned radial feature (NARF) for robust place recognition. A similar idea is also adopted by [3], where the difference is that the ORB point feature and corresponding visual validation paradigm are utilized. Instead of using the keypoint features, Dubé et al. [12] calculated features from the clustered point cloud segments. For the segments in a query scan, the obtained description vectors are sent to a classifier to identify matching segments in the database. In order to overcome the weak generalization of handcrafted features, a data-driven descriptor extraction network is proposed [13], where the classifier is also integrated. The aforementioned methods generally extract features based on parts of observation information. The absence of global representation is unfavorable for place recognition. In addition, the matching of a quantity of local features also possibly increases the computational amount.

9.2.2 Global-Descriptor-Based Methods

By taking the LiDAR observation as a whole, this solution generates the global description vector. Afterward, the place of a query observation is quickly retrieved based on the descriptor similarity. An ingenious implementation [14] is to aggregate local features and related spatiotemporal information into a robust and discriminative global descriptor. Differently, Röhling et al. [4] applied the histograms over simple global statistics of LiDAR scans such as height and range as descriptors for fast recognition. Finman et al. [15] explored object-based graphs to represent scenes, which can be regarded as a special global description. Then, the place recognition is performed according to graph matching. Due to the limitation of complexity, it is mainly used to handle small-scale environments. In M2DP (multi-view 2D projection) [16], multiple density signatures of points are generated by projecting the point cloud to different 2D planes, and the singular vectors of all planes are combined into a global descriptor. Given the compactness and global perspective of the bird's-eye-view [17] for point clouds, some methods have emerged recently, including scan context [5], LiDAR iris [18], and semantic scan context [19]. They discretize the bird's-eye-view with polar coordinates and encode the height or semantic information of points within sectors into a 2D matrix. The matrix distance is then considered as the evaluation criterion of similarity. These methods are efficient and less sensitive to the density of points.

With the development of deep learning, the methods generating the global descriptors by end-to-end learning attract more attention. Inspired by the success of NetVLAD in visual place recognition [11] and Chapter 5, Uy and Lee [6] combined PointNet and NetVLAD pooling to extract and aggregate point cloud features for obtaining the global descriptor. Further, Zhang and Xiao [20] introduced a point contextual attention network for feature adjustment. In NDT-Transformer [8], Zhou et al. additionally incorporated normal distribution transform information with transformer-based attention, enriching the learned descriptor with geometrical and contextual information. By quantizing the point cloud to a sparse tensor, MinkLoc3D [9] exploits a sparse-convolution-based 3D feature pyramid network and generalized mean pooling to produce a global descriptor. The aforementioned methods take a 3D representation of the point cloud as input. There are also 2D representation-based schemes. In LocNet [21], a minimalist network based on a handcrafted 2D histogram of range is designed to generate description vectors. FusionVLAD [22] integrates multi-view 2D projection representations to achieve viewpoint-free descriptor. Ma et al. [7] projected the point cloud to a range image and took advantage of the permutation equivariance of transformer as well as the permutation invariance of NetVLAD to achieve robustness to rotation. Despite obtaining good performance, it imposes restrictions on feature scale changes. Chen et al. [23] combined the range image and normal, intensity, and semantic images within a two-head-based siamese network to simultaneously calculate overlapping and yaw rotation. However, it is sensitive to the rotation of point clouds. Compared with previous studies, our network based on range image introduces scale-varied features while maintaining rotation invariance, achieving good performance.

9.3 LiDAR PLACE RECOGNITION BASED ON RANGE IMAGE AND COLUMN-SHIFT-INVARIANT ATTENTION

Figure 9.1 gives the architecture of the proposed LPR method, where a novel global descriptor extraction network is designed to generate a CSI descriptor for place recognition. Our network consists of a range image encoder, a feature aggregator, and a descriptor generator. The point cloud is converted to the range image $\mathcal{R}$, which is fed into the encoder, and then the column-shift-equivariant features are extracted. Based on the extracted feature map, the feature aggregator captures global contextual information with a CSI attention.

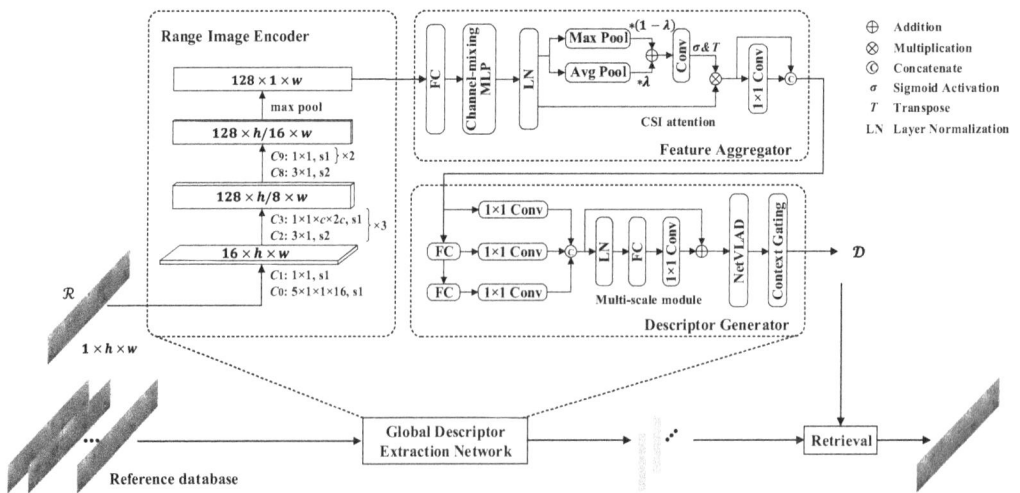

FIGURE 9.1 The pipeline of the proposed LiDAR-based place recognition method.

And the invariance to the yaw rotation is satisfied. On this basis, features at different scales are further learned by a multi-scale module to enhance discriminability, which is then encoded into a global description vector $\mathcal{D}$ for place recognition. With the extracted global descriptor of a query range image, a retrieval on the reference database is performed based on the descriptor similarity, and then the place recognition result is acquired.

9.3.1 Range Image Encoder

The point cloud $\mathcal{P}$ is first converted into the range image with sphere projection [23]. With the range image input $\mathcal{R}$, a feature encoder consisting of 11 convolution layers and a maximum pooling layer is tailored to extract column-shift-equivariant features. As illustrated in Figure 9.1, the kernel size of convolution layer C_k is labeled as $h_k \times w_k \times c_{in} \times c_{out}$, where k is the index of convolution layer. h_k, w_k, c_{in}, and c_{out} refer to height, width, number of input channels, and number of output channels, respectively. When c_{in} is equal to c_{out}, they are omitted in the expression of C_k for brevity. c and $2c$ of C_3 are only used to describe that the number of the output channels is twice as much as that of the input channels. All convolution layers adopt ReLU nonlinearity activation with biases disabled. The reduction of spatial dimensionality is handled by the downsampling based on strided convolution. In order to realize the equivariance to the cyclic column shift of range image, following [7], the downsampling operation is applied on the height dimension in the encoder. Moreover, the convolution also excludes the width dimension by setting the corresponding size of the filter kernel as 1 to handle the feature redundancy. Different from spatial downsampling implemented by strided convolution [24] that simultaneously contains channel upscaling, we refer to the depth-wise separate convolution [25] and decouple these two operations. Concretely, a strided convolution without channel upscaling and a 1×1 convolution with channel upscaling are applied sequentially, as shown in C_2 and C_3. This reduces the computational complexity and increases the nonlinearity of network. After the convolution and maximum pooling, a feature map with the size of $128 \times 1 \times w$ is outputted, where 128 refers to number of channels.

9.3.2 Feature Aggregator

Due to the absence of convolution on the width dimension, the information interaction in the encoder is restricted between the channel and height dimensions of the feature. This leads to the decreasing of discriminability. Besides, the extracted feature map is column-shift-equivariant. The lack of invariance constrains the change of feature map size. To facilitate information interaction and achieve invariance, a CSI attention module is presented. Before the feature map from the encoder is fed into the attention module, a fully connected layer without bias and a channel-mixing multiple layer perception (MLP) are employed in sequence to upscale the channel size and promote cross-channel information exchange. The channel-mixing MLP takes the form of an inverted bottleneck consisting of layer normalization, linear layers, GELU (Gaussian error linear unit) activation, and skip connection. Please refer to [26] for the details. The above-mentioned processing mainly works on the channel dimension, which is column-shift-equivariant, and the feature map size is transformed to $c_1 \times 1 \times w$, where c_1 refers to the channel dimension.

In the following, the attention module (see Figure 9.1) is designed, which mainly contains pooling and convolution. The input feature X is first processed with a layer normalization (LN) operation, whose output is copied and pooled into two vectors. Then, their weighted sum is transmitted to a 1D convolution with sigmoid activation for attention weight, which is given by

$$W(X) = \sigma\left(f_c\left(\lambda \cdot f_{ap}\left(f_{ln}(X)\right) + (1-\lambda) \cdot f_{mp}\left(f_{ln}(X)\right)\right)\right) \qquad (9.1)$$

where $f_{ln}(.)$ is layer normalization operation, $f_{ap}(\cdot)$ and $f_{mp}(\cdot)$ represent the global average pooling and maximum pooling along the feature channel, respectively. Their output sizes are both $1 \times w$. λ is a learnable scale weight. $f_c(\cdot)$ refers to a circular convolution (convolution with circular padding) and $\sigma(\cdot)$ means the sigmoid activation function. The 1D convolution upgrades the size of channel from 1 to c_2. Its kernel size is adaptively determined by $\left|(1+\log_2 w)/2\right|_{odd}$, where $|\cdot|_{odd}$ indicates the nearest odd number [27]. In deep learning, the implementation of convolution is equivalent to cross-correlation. Therefore, the formula (9.1) essentially calculates the correlations of convolution kernels and the weighted pooling output.

With the attention weight $W(X)$, the output of attention module is calculated, which achieves aggregation of the global spatial context information of X:

$$\mathcal{A}(X) = X \cdot \left(W(X)\right)^T \qquad (9.2)$$

Next, it is proved that the proposed attention module is CSI. According to the matrix theory, the cyclic column shift X_{cs} of X can be expressed as XC, where C is a special circulant matrix with each row being a standard unit vector and follows $C \cdot C^T = I$. The layer normalization f_{ln}, global pooling f_{ap}, f_{mp} and circular convolution f_c are equivariant to the cyclic column shift of input, satisfying $f(XC) = f(X)C$. Therefore,

$$\mathcal{A}(XC) = XC \cdot \left(W(XC)\right)^T$$

$$= XC \cdot \left(\sigma\left(f_c\left(\lambda \cdot f_{ap} f_{ln}(XC) + (1-\lambda) \cdot f_{mp} f_{ln}(XC)\right)\right)\right)^T$$

$$= XC \cdot \left(\sigma\left(f_c\left(\lambda \cdot f_{ap} f_{ln}(X) + (1-\lambda) \cdot f_{mp} f_{ln}(X)\right)C\right)\right)^T$$

$$= XC \cdot \left(\sigma\left(f_c\left(\lambda \cdot f_{ap} f_{ln}(X) + (1-\lambda) \cdot f_{mp} f_{ln}(X)\right)\right)C\right)^T \qquad (9.3)$$

$$= XC \cdot C^T \left(\sigma\left(f_c\left(\lambda \cdot f_{ap} f_{ln}(X) + (1-\lambda) \cdot f_{mp} f_{ln}(X)\right)\right)\right)^T$$

$$= X \cdot \left(W(X)\right)^T$$

$$= \mathcal{A}(X)$$

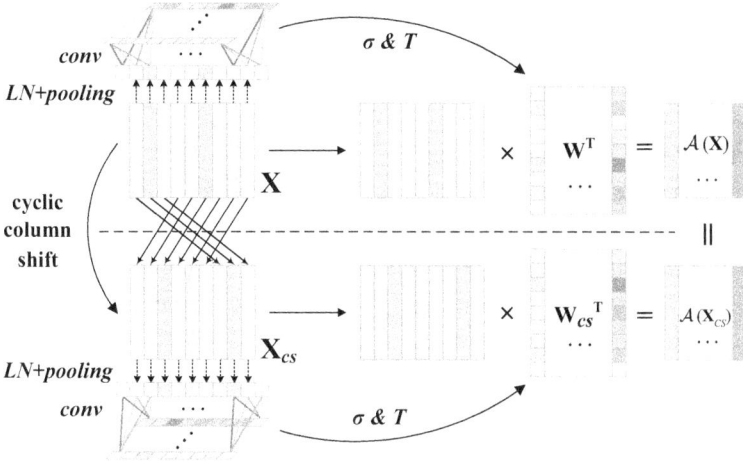

FIGURE 9.2   Illustration of the cyclic column shift invariance for the proposed attention module.

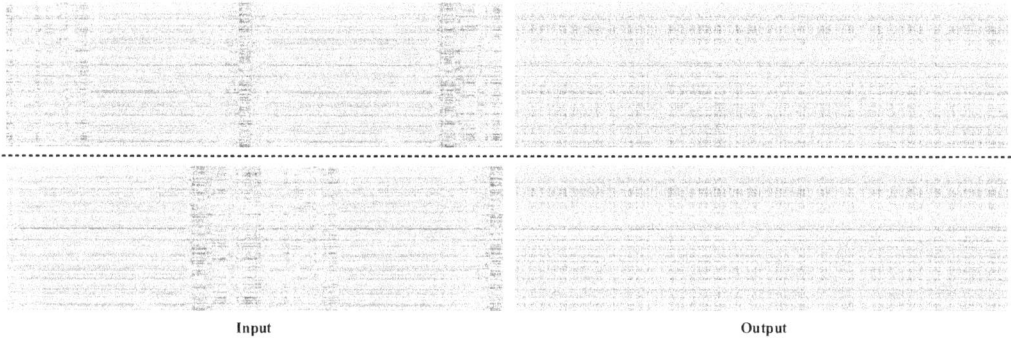

FIGURE 9.3   The invariance illustration of the proposed attention module. The inputs and outputs of the attention module for the first frame of KITTI 00 and its shifted version are given. The results indicate that the outputs are the same in spite of the cyclic column shift of input.

It is seen that the circulant matrix C in XC is neutralized by the matrix C^T, resulting in the cyclic column shift invariance. Figure 9.2 presents an example of invariance considering a single pooling. It can be seen that the output of the attention block remains unchanged. Take the first frame of sequence 00 in the KITTI dataset and its shifted version as an example. The results of our attention module are illustrated in Figure 9.3. It is seen that their outputs are the same in spite of the cyclic column shift of input.

The dotted arrows and straight lines represent pooling and convolution operations, respectively. Because the convolution enables the synchronous column shift of the attention weight $W(X)$ and X, $\mathcal{A}(X)_{*,j} = \sum_i X_{*,i} \cdot \left(W^T\right)_{i,j} = \sum_i (X_{cs})_{*,i} \cdot \left(W_{cs}^T\right)_{i,j} = \mathcal{A}(X_{cs})_{*,j}$.

Self-attention [10] linearly transforms the feature embedding into query, key, and value matrixes. It can be noticed that our attention works on a similar principle as self-attention, and the pooling output, the convolution kernels, and X may be considered as query, key, and value, respectively. Different from self-attention, whose key changes dynamically, the key in our module is constant after training. This enables the synchronous shift of attention maps and then achieves the invariance to the cyclic column shift. The output of the attention module is sent to a linear transformation for better feature aggregation.

9.3.3 Descriptor Generator

In various deep learning networks, multi-scale strategies with feature dimensionality reduction have shown great potential for performance improvement. However, the requirement of constant dimension in width for the column shift equivariance hinders the multi-scale application, which limits the precision of place recognition. Benefiting from the invariance of the proposed attention module, the obstruction is removed. Herein, a multi-scale module is designed to enhance the feature discriminability. It consists of downsampling and feature mixing. Two fully connected layers are cascaded to realize the progressive reduction of spatial dimension. Three convolutions are applied to different scales, and their outputs are concatenated along width and then mixed. The mixing operation takes the form of residual including layer normalization, FC, and 1×1 convolution, where the latter two act on spatial dimension and channel dimension, respectively. The output after mixing is fed into a NetVLAD layer [1] to aggregate the learned feature into the VLAD bag-of-words vectors. Considering the retrieval complexity of place recognition, the generated vector is compressed into a compact form with a fully connected layer. After that, in order to enhance the discriminability, a context gating unit is attached. Then L2-normalization is employed and the final global descriptor $\mathcal{D}$ with the size of d_g is generated.

9.3.4 Loss Function

The proposed global descriptor extraction network is trained end-to-end. Generally, as the network gets deeper, the model is more difficult to be trained. Besides the commonly used lazy triplet loss [6,22] applied to the descriptor $\mathcal{D}$, another loss of the same type is introduced to facilitate the learning of features, which works on the vector $\mathcal{V}$ generated by applying global max-pooling on the output of the encoder along the width dimension. Given a training tuple $\left(\mathcal{P}_a, \{\mathcal{P}_{p_i} \mid 1 \le i \le N_p\}, \{\mathcal{P}_{n_j} \mid 1 \le j \le N_n\}\right)$, we denote the corresponding pooling vectors and global descriptors as $\left(\mathcal{V}_a, \{\mathcal{V}_{p_i} \mid 1 \le i \le N_p\}, \{\mathcal{V}_{n_j} \mid 1 \le j \le N_n\}\right)$ and $\left(\mathcal{D}_a, \{\mathcal{D}_{p_i} \mid 1 \le i \le N_p\}, \{\mathcal{D}_{n_j} \mid 1 \le j \le N_n\}\right)$, where N_p and N_n are the number of positive and negative samples in the tuple; the symbol $\{\cdot\}$ represents a set; the subscripts a, p, and n refer to anchor, positive and negative samples, respectively. The training loss can be expressed as follows:

$$\mathcal{L} = \sum_j \left\{ \gamma \cdot \left[\alpha - d\left(\mathcal{V}_a, \mathcal{V}_{n_j}\right) + \max_{1 \le i \le N_p} d\left(\mathcal{V}_a, \mathcal{V}_{p_i}\right) \right]_+ + (1-\gamma) \cdot \left[\beta - d\left(\mathcal{D}_a, \mathcal{D}_{n_j}\right) + \max_{1 \le i \le N_p} d\left(\mathcal{D}_a, \mathcal{D}_{p_i}\right) \right]_+ \right\}$$

(9.4)

where $d(\cdot)$ refers to the distance metric function and the squared Euclidean distance is utilized. α and β are the predefined margins. $[\cdot]_+$ denotes the hinge loss. γ is a dynamic weight to balance the two loss terms.

9.3.5 LiDAR Place Recognition via Descriptor-Based Similarity Retrieval

With the learned global descriptors, the similarity of two descriptors is utilized to evaluate the probability of whether the corresponding places are the same. Specifically, we precomputed the corresponding global descriptors for all point clouds in the reference database, which constitute the descriptor set. Given a query point cloud, it is first mapped into a compressed global descriptor by the proposed network. The similarity between the query descriptor and each reference descriptor is calculated, where the Euclidean distance is taken as similarity metric. Finally, the candidate-recognized places are selected according to the similarity ranking. Algorithm 9.1 describes the retrieval process, where $\mathcal{T}_P$ and $\mathcal{T}_D$ refer to the reference database and corresponding descriptor set.

Algorithm 9.1 Retrieval-Based LiDAR Place Recognition

Input: query point cloud $\mathcal{P}$, CSI network $f(\cdot)$, reference database $\mathcal{T}_P = \{\mathcal{P}_1, \ldots, \mathcal{P}_n, \ldots\}$ and corresponding descriptor set $\mathcal{T}_D = \{\mathcal{D}_1, \ldots, \mathcal{D}_n, \ldots\}$.

Output: top-k candidates $\mathcal{O}$.

1   Convert $\mathcal{P}$ to the range image $\mathcal{R}$ by sphere projection;
2   $\mathcal{D} = f(\mathcal{R})$;
3   **for** l from 1 to $|\mathcal{T}_D|$ **do**
4   Calculate similarity s_l between $\mathcal{D}$ and $\mathcal{D}_l$;
5   **end for**
6   Sort $\{s_1, \ldots, s_n, \ldots\}$ in a descending order and obtain $\{s_{a_1}, \ldots, s_{a_n}, \ldots\}$;
7   $\mathcal{O} = \{\mathcal{P}_{a_i} \mid 1 \le i \le k\}, \mathcal{P}_{a_i} \in \mathcal{T}_P$;
8   **return**

9.4 EXPERIMENTS

We conduct experiments on KITTI [28], Ford Campus [29], and NCLT [30] datasets to evaluate the performance of the proposed method termed as CSINet. The point clouds of the first two datasets are collected with a Velodyne HDL-64E LiDAR mounted on a vehicle, which covers multiple town scenes. The NCLT dataset is acquired by a Segway robot equipped with a Velodyne HDL-32E LiDAR in a campus across different seasons. For the range image projected from point cloud, the height is set to the corresponding scan lines of LiDAR, and the width is fixed to 900. The network is implemented with PyTorch framework and trained on four NVIDIA RTX 1080Ti GPUs for 50 epochs. The batch size of the triplet tuple is 32, where the number of positive and negative samples in a tuple are both set to 6. The initial learning rate is 1e-4, which decays by 0.8 every five epochs. In the loss function, the margins α and β are 0.25 and 0.5, respectively. For the weight γ, its initial value is set to 0.5, and the corresponding decay rate and step size are 0.8 and 10 epochs.

9.4.1 Accuracy Evaluation

In order to demonstrate the proposed approach, comparative experiments are carried out. Firstly, we test the place recognition performance of CSINet on the NCLT dataset. The network is trained on the oldest sequence 2012-01-08 and evaluated on other sequences 2012-02-05, 2012-06-15, 2013-02-23, and 2013-04-05, where the name of the sequence identifies its collection time. Given an anchor scan, its positive and negative samples are determined by the Euclidean distance between their ground-truth poses. During training, the thresholds for positive and negative pairs are $10\,m$ (<) and $50\,m$ (>), respectively. For testing, place recognition is regarded as correct when the distance between retrieval and query frames is less than $15\,m$ [7]. When evaluating, we retrieve candidate places from sequence 2012-01-08 for a query scan in other sequences, and the average top-k recalls (AR@1, AR@5, and AR@20) are considered as evaluation metrics. The comparison methods include PointNetVLAD [6], MinkLoc3D [9], OverlapTransformer [7], HRS [31], SeqLPD [32], CIMV [33], SeqNet [34], and SeqOT [35]. The first three approaches and our CSINet are based on a single scan, and the rest leverages consecutive multiple scans. Particularly, HRS is a glued method combining a hashing-based retrieval strategy and features generated by OverlapTransformer. CIMV and SeqNet are visual place recognition methods that directly operate on range image. Table 9.1 presents the results of different methods, where the results of the compared methods are from [35] and the best results are labeled in bold. This verifies the proposed method.

The proposed CSINet is also verified on the KITTI and Ford Campus datasets by the loop closure detection task. We take the old scans before the query scan as reference scans. Also, the past 100 adjacent scans of query scan are excluded to ensure the time interval between loop closure scans. The KITTI dataset contains 11 sequences with ground-truth poses, where the last eight sequences are utilized for training. The validation and testing of detection performance are operated on sequences 02 and 00, respectively. To assess the generalization of our method, we directly test on the sequence 00 of the Ford Campus using the model trained on the KITTI dataset without additional training. For valid loop closure, the detected frame should have a high overlap with the query frame, which could not be guaranteed only according to the distance metric. Hence, we take the range-image-based overlap between two scans as the criterion to generate more suitable positive and negative samples for supervised training. The scan whose overlap with the anchor is greater than 0.3 is regarded as a positive sample [7], otherwise, it is a negative sample. In testing, the same threshold is employed. We compare CSINet with existing LiDAR-based place recognition methods including Histogram [4], Scan Context [5], LiDAR Iris [18], FreSCo [36], PointNetVLAD [6], OverlapNet [23], NDT-Transformer-P [8], MinkLoc3D [9], and OverlapTransformer [7]. The first four are traditional methods and the last five are based on deep learning. The metrics AUC, average recall 1 (AR@1), and average recall 1% (AR@1%) are utilized for quantitative evaluation. The data of FreSCo [36] is obtained with the corresponding open source code, and the results of other comparison methods are from [7]. Table 9.2 shows the comparison results on sequences 00 of KITTI and Ford Campus datasets. It is observed that the proposed CSINet performs well on the KITTI 00. For the Ford 00, the results of our method also show good generalization.

TABLE 9.1 Comparison of Different Methods for Place Recognition on the NCLT Dataset

Method		2012-02-05			2012-06-15			2013-02-23			2013-04-05		
		AR@1	AR@5	AR@20	AR@1	AR@5	AR@20	AR@1	AR@5	AR@20	AR@1	AR@5	AR@20
Multi-Scan-Based	HRS	0.869	0.925	0.955	0.624	0.716	0.821	0.557	0.669	0.814	0.498	0.643	0.752
	CIMV	0.871	0.925	0.957	0.642	0.730	0.852	0.564	0.707	0.835	0.527	0.663	0.797
	SeqNet	0.889	0.933	0.960	0.645	0.745	0.859	0.569	0.725	0.847	0.517	0.674	0.801
	SeqLPD	0.873	0.928	0.952	0.663	0.791	0.884	0.658	0.713	0.847	0.582	0.719	**0.835**
	SeqOT	0.917	0.947	**0.968**	**0.762**	**0.844**	**0.899**	**0.691**	0.723	**0.874**	0.639	0.724	0.826
Single-Scan-Based	PointNetVLAD	0.746	0.823	0.875	0.612	0.720	0.782	0.469	0.604	0.719	0.449	0.576	0.683
	MinkLoc3D	0.802	0.864	0.926	0.630	0.685	0.774	0.507	0.616	0.751	0.482	0.587	0.685
	OverlapTransformer	0.861	0.899	0.930	0.639	0.697	0.780	0.536	0.645	0.764	0.496	0.603	0.715
	CSINet	**0.929**	**0.948**	0.958	0.706	0.774	0.827	0.673	**0.762**	0.823	**0.645**	**0.728**	0.804

TABLE 9.2   Comparison of Loop Closure Detection Performance for Different Methods on the Sequences 00 of KITTI and Ford Campus Datasets

Sequence	Method	AUC	AR@1	AR@1%
KITTI 00	Histogram	0.826	0.738	0.871
	Scan Context	0.836	0.820	0.869
	LiDAR Iris	0.843	0.835	0.877
	FreSCo	0.891	0.883	0.974
	PointNetVLAD	0.856	0.776	0.845
	OverlapNet	0.867	0.816	0.908
	NDT-Transformer-P	0.855	0.802	0.869
	MinkLoc3D	0.894	0.876	0.920
	OverlapTransformer	0.907	0.906	0.964
	CSINet	**0.914**	**0.941**	**0.979**
Ford 00	Histogram	0.841	0.812	0.897
	Scan Context	0.903	0.878	0.958
	LiDAR Iris	0.907	0.849	0.937
	FreSCo	**0.929**	0.892	0.967
	PointNetVLAD	0.872	0.862	0.938
	OverlapNet	0.854	0.857	0.932
	NDT-Transformer-P	0.835	0.900	0.927
	MinkLoc3D	0.871	0.878	0.942
	OverlapTransformer	0.923	0.914	0.954
	CSINet	0.916	**0.941**	**0.984**

The results of the proposed CSINet on Ford 00 are obtained directly using the corresponding model trained on the KITTI dataset.

9.4.2 Robustness Test

To demonstrate the robustness of our method to viewpoint change, we focus on image shift. Meanwhile, Scan Context [5], LiDAR Iris [18], and OverlapTransformer [7] are chosen for comparison on average recall 1(AR@1). The sequence 00 of the KITTI dataset is chosen and the old frames constitute the database for retrieval. In the experiment I, the query images of different methods are cyclically shifted along the column. And the experiment II directly rotates the query point clouds with random yaw. The results of the experiments I and II are presented in Figure 9.4a and b, respectively. One can conclude from Figure 9.4a that the AR@1 of all methods remains constant, which implies the invariance to the cyclic column shift of the image. As for the rotation of the point cloud, interference appears due to the discretization error during point cloud projection. From Figure 9.4b, the AR@1 curves of all methods fluctuate. And the precision change of our method and OverlapTransformer is smaller. The robustness of our CSINet to the point cloud rotation is proved.

In addition, the examples of loop closure detection and place recognition of the proposed method are provided in Figure 9.5. Figure 9.5a describes the result of loop closure for a query scan on the Ford 00. The results in NCLT dataset are given in Figure 9.5b and c for query scans from sequences 2012-06-15 and 2013-04-05, respectively, where sequence 2012-01-08 is regarded as reference database. In spite of the difference in orientation or position, the top-1 candidate frames are successfully retrieved.

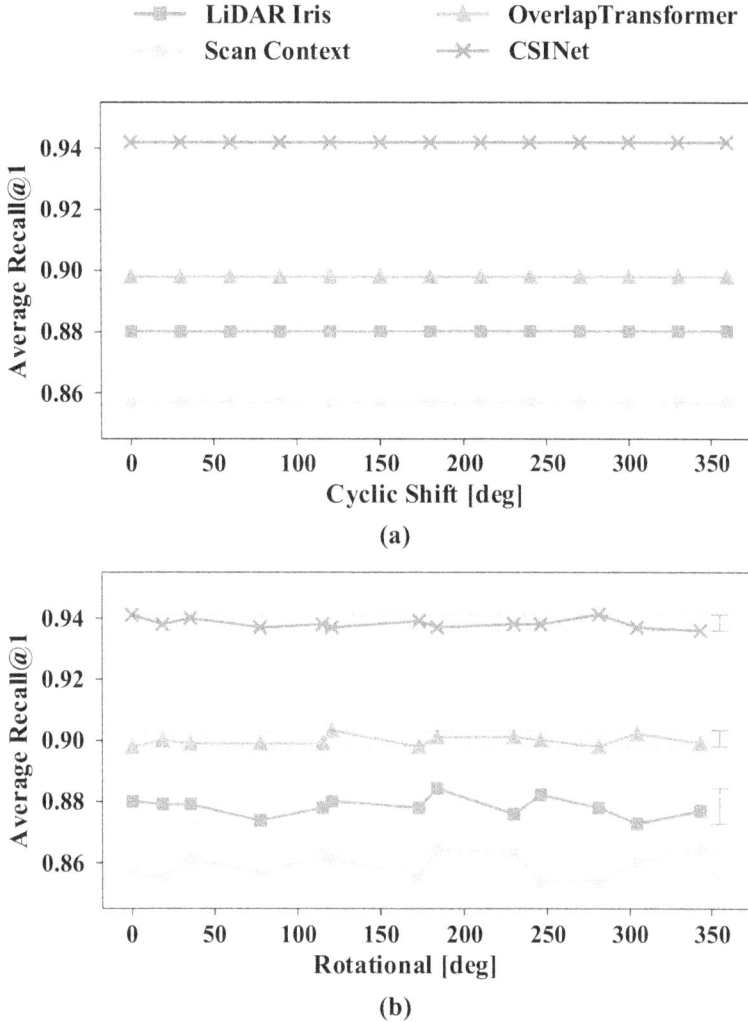

FIGURE 9.4 Robustness test on sequence 00 of the KITTI dataset. (a) Cyclic shift of range images along the column. (b) Point cloud rotation around the yaw-axis.

9.4.3 Ablation Study

In this section, the effects of different components of the proposed method are investigated on KITTI 00 and Ford 00, where the mixed loss, CSI attention, and multi-scale features are involved. CSINet-I is the simplest version that excludes all parts. CSINet-II and CSINet-III introduce mixed loss and CSI attention, respectively, while CSINet-IV includes both components. Table 9.3 presents the results of different variants in terms of AR@1 and AR@1%. In comparison with CSINet-I, the accuracy of CSINet-II and CSINet-III is improved with the aid of mixed loss or CSI attention. From the results of CSINet-II, CSINet-III, and CSINet-IV, the performance of CSINet-IV fluctuates, such as the AR@1% result in KITTI 00 and the AR@1 result in Ford 00. According to the training loss, the training effect of the network is closer to maximize the AR@1 instead of AR@1% due to the limited numbers

(a)

(b)

(c)

FIGURE 9.5   The illustration of loop closure detection and place recognition of our method on Ford Campus and NCLT datasets, where the query, top-1 retrieval result, and trajectory are given. The arrows in the subfigure of trajectory indicate the motion trend. (a) Loop closure detection result for a query scan on the Ford 00. (b) Result of place recognition for a query scan from the sequence 2012-06-15 on the sequence 2012-01-08 (reference database). (c) Result of place recognition for a query scan from the sequence 2013-04-05 on the sequence 2012-01-08 (reference database).

TABLE 9.3   Comparison of Different Variants of Our Method on KITTI 00 and Ford 00

Method	Mixed Loss	CSI Attention	Multi-Scale Features	KITTI 00		Ford 00	
				AR@1	AR@1%	AR@1	AR@1%
CSINet-I	×	×	×	0.900	0.950	0.923	0.972
CSINet-II	√	×	×	0.912	0.975	0.933	0.974
CSINet-III	×	√	×	0.912	0.976	0.932	0.972
CSINet-IV	√	√	×	0.928	0.974	0.927	0.980
CSINet	√	√	√	**0.941**	**0.979**	**0.941**	**0.984**

The results of Ford 00 are obtained directly using the corresponding models trained on the KITTI dataset.

of positive and negative samples in a training tuple. This also explains that the AR@1% in KITTI 00 is not always stable. It is noted that the results in Ford 00 are directly obtained using the model trained on the KITTI dataset. It may make the network prefer to data from KITTI, which possibly weakens the adaptability of CSINet-IV to Ford. Finally, the introduction of multi-scale module enriches the features and the CSINet attains the best results in both KITTI 00 and Ford 00.

Besides, we also analyze the influence of the output dimension of global descriptor on recognition precision. Table 9.4 presents the results of the proposed method on the NCLT dataset in terms of AR@1 and AR@5 with the output dimensions of 128, 256, and 512. Overall, the output dimension of 256 brings in better accuracy and it is adopted in our CSINet.

9.4.4 Efficiency Analysis

In this section, the efficiency of the proposed method is analyzed. Firstly, we compare the time of generating global descriptor, where the inferences of deep networks are accomplished based on the LibTorch library. The first frame of sequence 00 in the KITTI dataset is adopted for 2,000 tests, and the average processing time of global descriptor generation for different methods is shown in Table 9.5. All methods in Table 9.5 are executed on the same PC with an Intel i7–9750H CPU and an NVIDIA RTX 2080 GPU, where the hand-crafted methods run only using CPU. It can be seen that our method can run at almost 397 Hz, which is slower than OverlapTransformer due to the fact that our method leverages

TABLE 9.4   Performance Comparison of Our Method with Different Descriptor Output Dimensions on the NCLT Dataset

	$d_g = 128$		$d_g = 256$		$d_g = 512$	
Sequence	AR@1	AR@5	AR@1	AR@5	AR@1	AR@5
2012-02-05	0.927	0.947	**0.929**	0.948	0.927	**0.950**
2012-06-15	0.700	0.770	**0.706**	0.774	0.700	**0.775**
2013-02-23	0.670	0.751	0.673	**0.762**	**0.680**	0.752
2013-04-05	0.616	0.706	**0.645**	**0.728**	0.621	0.702

TABLE 9.5   Average Processing Time of Global Descriptor Generation

	Methods	Average Processing Time Per Frame (ms)
Handcrafted	Scan Context	3.30
	LiDAR Iris	9.12
Learning-Based	PointNetVLAD	3.96
	OverlapNet	3.64
	OverlapTransformer	2.05
	NDT-Transformer	8.02
	CSINet	2.52

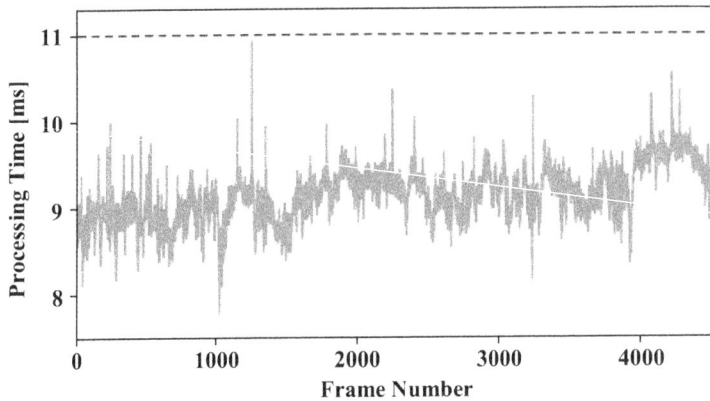

FIGURE 9.6 The processing time of loop closure detection for each frame in KITTI sequence 00.

more complicated architecture to improve precision. Further, we evaluate the runtime of online loop closure detection, where the point cloud preprocessing and the retrieval are also included, and Faiss library is utilized for quick search. We dynamically simulate the process with sequence 00 of the KITTI dataset, and the top 10 candidates for every frame are retrieved. The corresponding runtime is illustrated in Figure 9.6, where the retrieval is not executed for the first 100 scans. The average processing time is about 9.2 ms, and the efficiency of the proposed method is fine.

9.5 CONCLUSION

This chapter proposes a global descriptor extraction network based on the point cloud range image for LPR. Aiming at the difficulty of applying multi-scale features, a cyclic-column-shift-invariant attention is specially designed. This improves the robustness of place recognition to viewpoint changes, and enables the subsequent multi-scale features. Then, a multi-scale feature module is presented for enriching features with the adaptation to spatial discontinuity of input features. As a result, the discriminability of global descriptors is enhanced, which promotes the place recognition performance. The experiment results on the datasets demonstrate the effectiveness and generalization of the proposed method.

REFERENCES

[1] Guo, J., Borges, P. V. K., Park, C., & Gawel, A. (2019). Local descriptor for robust place recognition using LiDAR intensity. *IEEE Robotics and Automation Letters*, 4(2), 1470–1477.

[2] Steder, B., Ruhnke, M., Grzonka, S., & Burgard, W. (2011). Place recognition in 3D scans using a combination of bag of words and point feature based relative pose estimation. In *Proceedings of the IEEE/RSJ International Conference on Intelligent Robots and Systems*, San Francisco, CA, USA (pp. 1249–1255).

[3] Shan, T., Englot, B., Duarte, F., Ratti, C., & Rus, D. (2021). Robust place recognition using an imaging LiDAR. In *Proceedings of the IEEE International Conference on Robotics and Automation*, Xi'an, China (pp. 5469–5475).

[4] Röhling, T., Mack, J., & Schulz, D. (2015). A fast histogram-based similarity measure for detecting loop closures in 3-D LiDAR data. In *Proceedings of the IEEE/RSJ International Conference on Intelligent Robots and Systems*, Hamburg, Germany (pp. 736–741).

[5] Kim, G., & Kim, A. (2018). Scan context: Egocentric spatial descriptor for place recognition within 3D point cloud map. In *Proceedings of the IEEE/RSJ International Conference on Intelligent Robots and Systems*, Madrid, Spain (pp. 4802–4809).

[6] Uy, M. A., & Lee, G. H. (2018). PointNetVLAD: Deep point cloud based retrieval for large-scale place recognition. In *Proceedings of the IEEE Conference on Computer Vision and Pattern Recognition*, Salt Lake City, UT, USA (pp. 4470–4479).

[7] Ma, J., Zhang, J., Xu, J., Ai, R., Gu, W., & Chen, X. (2022). OverlapTransformer: An efficient and yaw-angle-invariant transformer network for LiDAR-based place recognition. *IEEE Robotics and Automation Letters*, 7(3), 6958–6965.

[8] Zhou, Z., Zhao, C., Adolfsson, D., Su, S., Gao, Y., Duckett, T., & Sun, L. (2021). NDT-Transformer: Large-Scale 3D point cloud localisation using the normal distribution transform representation. In *Proceedings of the IEEE International Conference on Robotics and Automation*, Xi'an, China (pp. 5654–5660).

[9] Komorowski, J. (2021). MinkLoc3D: Point cloud based large-scale place recognition. In *Proceedings of the IEEE Winter Conference on Applications of Computer Vision*, Waikoloa, HI, USA (pp. 1789–1798).

[10] Vaswani, A., Shazeer, N., Parmar, N., Uszkoreit, J., Jones, L., Gomez, A. N., Kaiser, Ł., & Polosukhin, I. (2017). Attention is all you need. In *Advances in Neural Information Processing Systems*, Long Beach, CA, USA (pp. 1–11).

[11] Arandjelovic, R., Gronat, P., Torii, A., Pajdla, T., & Sivic, J. (2016). NetVLAD: CNN architecture for weakly supervised place recognition. In *Proceedings of the IEEE Conference on Computer Vision and Pattern Recognition*, Las Vegas, NV, USA (pp. 5297–5307).

[12] Dubé, R., Dugas, D., Stumm, E., Nieto, J., Siegwart, R., & Cadena, C. (2017). SegMatch: Segment based place recognition in 3D point clouds. In *Proceedings of the IEEE International Conference on Robotics and Automation*, Singapore (pp. 5266–5272).

[13] Dubé, R., Cramariuc, A., Dugas, D., Nieto, J., Siegwart, R., & Cadena, C. (2018). SegMap: 3D segment mapping using data-driven descriptors. arXiv preprint arXiv:1804.09557.

[14] Vidanapathirana, K., Moghadam, P., Harwood, B., Zhao, M., Sridharan, S., & Fookes, C. (2021). Locus: LiDAR-based place recognition using spatiotemporal higher-order pooling. In *Proceedings of the IEEE International Conference on Robotics and Automation*, Xi'an, China (pp. 5075–5081).

[15] Finman, R., Paull, L., & Leonard, J. J. (2015). Toward object-based place recognition in dense RGB-D maps. In *ICRA Workshop Visual Place Recognition in Changing Environments*, 76, 480.

[16] He, L., Wang, X., & Zhang, H. (2016). M2DP: A novel 3D point cloud descriptor and its application in loop closure detection. In *Proceedings of the IEEE/RSJ International Conference on Intelligent Robots and Systems*, Daejeon, Korea (South) (pp. 231–237).

[17] Wu, Z., Gan, Y., Li, X., Wu, Y., Wang, X., Xu, T., & Wang, F. (2023). Surround-view fisheye BEV-perception for valet parking: Dataset, baseline and distortion-insensitive multi-task framework. *IEEE Transactions on Intelligent Vehicles*, 8(3), 2037–2048.

[18] Wang, Y., Sun, Z., Xu, C. Z., Sarma, S. E., Yang, J., & Kong, H. (2020). LiDAR iris for loop-closure detection. In *Proceedings of the IEEE/RSJ International Conference on Intelligent Robots and Systems*, Las Vegas, NV, USA (pp. 5769–5775).

[19] Li, L., Kong, X., Zhao, X., Huang, T., Li, W., Wen, F., Zhang, H., & Liu, Y. (2021). SSC: Semantic scan context for large-scale place recognition. In *Proceedings of the IEEE/RSJ International Conference on Intelligent Robots and Systems*, Prague, Czech Republic (pp. 2092–2099).

[20] Zhang, W., & Xiao, C. (2019). PCAN: 3D attention map learning using contextual information for point cloud based retrieval. In *Proceedings of the IEEE/CVF Conference on Computer Vision and Pattern Recognition*, Long Beach, CA, USA (pp. 12428–12437).

[21] Yin, H., Tang, L., Ding, X., Wang, Y., & Xiong, R. (2018). LocNet: Global localization in 3D point clouds for mobile vehicles. In *Proceedings of IEEE Intelligent Vehicles Symposium*, Changshu, China, Changshu, China (pp. 728–733).

[22] Yin, P., Xu, L., Zhang, J., & Choset, H. (2021). FusionVLAD: A multi-view deep fusion networks for viewpoint-free 3D place recognition. *IEEE Robotics and Automation Letters*, 6(2), 2304–2310.

[23] Chen, X., Läbe, T., Milioto, A., Röhling, T., Behley, J., & Stachniss, C. (2022). OverlapNet: A siamese network for computing LiDAR scan similarity with applications to loop closing and localization. *Autonomous Robots*, 46, 61–81.

[24] He, K., Zhang, X., Ren, S., & Sun, J. (2016). Deep residual learning for image recognition. In *Proceedings of the IEEE Conference on Computer Vision and Pattern Recognition*, Las Vegas, NV, USA (pp. 770–778).

[25] Howard, A. G., Zhu, M., Chen, B., Kalenichenko, D., Wang, W., Weyand, T., Andreetto, M., & Adam, H. (2017). MobileNets: Efficient convolutional neural networks for mobile vision applications. arXiv preprint arXiv:1704.04861.

[26] Tolstikhin, I. O., Houlsby, N., Kolesnikov, A., Beyer, L., Zhai, X., Unterthiner, T., Yung, J., Steiner, A., Keysers, D., Uszkoreit, J., Lucic, M., & Dosovitskiy, A. (2021). MLP-Mixer: An all-MLP architecture for vision. In *Advances in Neural Information Processing Systems* (pp. 24261–24272).

[27] Wang, Q., Wu, B., Zhu, P., Li, P., Zuo, W., & Hu, Q. (2020). ECA-Net: Efficient channel attention for deep convolutional neural networks. In *Proceedings of the IEEE/CVF Conference on Computer Vision and Pattern Recognition*, Seattle, WA, USA (pp. 11531–11539).

[28] Geiger, A., Lenz, P., & Urtasun, R. (2012). Are we ready for autonomous driving? The KITTI vision benchmark suite. In *Proceedings of the IEEE Conference on Computer Vision and Pattern Recognition*, Providence, RI, USA (pp. 3354–3361).

[29] Pandey, G., McBride, J. R., & Eustice, R. M. (2011). Ford campus vision and LiDAR data set. *The International Journal of Robotics Research*, 30(13), 1543–1552.

[30] Carlevaris-Bianco, N., Ushani, A. K., & Eustice, R. M. (2016). University of Michigan North Campus long-term vision and LiDAR dataset. *The International Journal of Robotics Research*, 35(9), 1023–1035.

[31] Vysotska, O., & Stachniss, C. (2017). Relocalization under substantial appearance changes using hashing. In *Proceedings of the IROS Workshop on Planning, Perception and Navigation for Intelligent Vehicles*, Vancouver, Canada (pp. 1–7).

[32] Liu, Z., Suo, C., Zhou, S., Xu, F., Wei, H., Chen, W., Wang, H., Liang, X., & Liu, Y. H. (2019). SeqLPD: Sequence matching enhanced loop-closure detection based on large-scale point cloud description for self-driving vehicles. In *Proceedings of the IEEE/RSJ International Conference on Intelligent Robots and Systems*, Macau, China (pp. 1218–1223).

[33] Facil, J. M., Olid, D., Montesano, L., & Civera, J. (2019). Condition-invariant multi-view place recognition. arXiv preprint arXiv:1902.09516.

[34] Garg, S., & Milford, M. (2021). SeqNet: Learning descriptors for sequence-based hierarchical place recognition. *IEEE Robotics and Automation Letters*, 6(3), 4305–4312.

[35] Ma, J., Chen, X., Xu, J., & Xiong, G. (2023). SeqOT: A spatial-temporal transformer network for place recognition using sequential LiDAR data. *IEEE Transactions on Industrial Electronics*, 70(8), 8225–8234.

[36] Fan, Y., Du, X., Luo, L., & Shen, J. (2022). FreSCo: Frequency-domain scan context for LiDAR-based place recognition with translation and rotation invariance. In *International Conference on Control, Automation, Robotics and Vision*, Singapore (pp. 576–583).

CHAPTER 10

Summary and Outlook

THIS BOOK FOCUSES ON localization and mapping technologies of autonomous mobile robots, which involves visual localization and mapping, visual relocalization, LiDAR localization and mapping as well as place recognition, etc. The introduction lays a solid foundation by highlighting the significance of localization and mapping in autonomous mobile robot domain.

Given the visual images captured online within an unknown environment, a visual SLAM method based on point and object semantic features, called PO-SLAM, is proposed to provide additional semantic constraints for optimizing camera poses and maps, thereby improving localization accuracy. An object segmentation strategy that combines 2D object detection and depth maps is designed to accelerate segmentation speed and facilitate data association between point features and object features within the same frame. Furthermore, feature points and objects between two frames are associated, and their semantic features and association results are fed into bundle adjustment (BA) optimization. In addition to the point-to-point error, a point-to-object error constraint is introduced, ensuring that the 3D points' projections in the current frame fall within their corresponding object detection boxes. Additionally, an object-to-object error constraint is constructed, maintaining the direction and length of line segments between any two static objects within the current frame's field of view. These constraints enable more accurate localization. The effectiveness of the proposed method is verified through experiments.

With the assistance of a previously built environment map, the robot can achieve more precise global relocalization. According to the storage format of the offline map, this book introduces two kinds of visual relocalization methods. Firstly, a camera relocalization method based on a scene coordinate regression network, called SFT-CR, is proposed. A scene coordinate regression network based on spatial feature transformation is designed, incorporating the CoordConv operation into the feature extraction module to provide additional positional information, enhancing the discriminability of features in low-texture regions. Meanwhile, under the guidance of the global information of the image, a position-wise local transformation is applied to the source feature map obtained from the feature extraction module, guiding the transformation of features from different viewpoints into a common space and achieving the decoupling of features and viewpoints. A maximum likelihood-based loss function is designed to train the network to simultaneously learn the 3D coordinates corresponding to pixels in the input image and their uncertainty. During the testing phase, the predicted 2D-3D point pairs are filtered based on the uncertainty output by the network, enabling the rapid and accurate computation of the camera's 6D pose. The effectiveness of the method is demonstrated through experiments on datasets. Then, a visual place recognition method based on multi-task learning, called MTA, is proposed to facilitate relocalization under the large-scale scene. A classification task containing a binary classification network and binary classification loss is constructed and combined with the existing triplet ranking task to jointly train the global feature extraction network. This approach improves the compactness of feature distributions at nearby locations and enhances the distinguishability of features at distant locations, thereby improving the model's generalization performance. In addition, an attention module is designed to capture flexible multi-scale contextual information while adjusting the distribution of convolutional features to address the effects of environmental changes. This allows the network to focus on regions that are useful for place recognition. Experiments show that the proposed method effectively extracts discriminative global image features, significantly improving the recall rate for place recognition.

With respect to the real application of different visual localization methods, a software architecture for service robot localization is proposed to integrate the above visual localization methods. Firstly, an offline hybrid map construction method combining explicit and implicit submaps is proposed. The explicit submap of the environment is built by integrating PO-SLAM and the global feature extraction networks from MTA, while the implicit submap is represented as the scene coordinate regression network, enabling the scene structure to be implicitly stored in the network parameters. A reference coordinate system is also defined to unify the submaps of different regions. During the online robot localization, when PO-SLAM tracking fails, the robot recovers its pose by combining the global features of the keyframes and the scene coordinate regression network, prioritizing the submap of the region from the previous moment. Simultaneously, unsuitable explicit submaps are quickly filtered based on the global feature similarity between the current image and the explicit submaps. Navigation experiments in indoor office environments demonstrate that the proposed localization framework enables continuous and stable localization of the robot in large-scale environments.

Aiming at the contradiction between the accuracy and efficiency in distribution-based 3D LiDAR odometry, a LiDAR odometry based on tightly associated distribution and maximum likelihood estimation is proposed. A sparse data association from source distribution to target distribution is designed. For each pair of associated distributions, the union of corresponding point sets is utilized to represent the sampling distribution, which associates the matched distributions tightly and thereby improves the representation of local structures. On this basis, a cost function that takes the point-to-reference distribution distances as constraints is constructed, and a decoupling strategy is presented. Through the pre-calculation of distribution parameters, the optimization complexity is decoupled from the number of points. As a result, the accurate pose and real-time performance are both obtained. Meanwhile, the aforementioned data association and optimization scheme is generalized to the situation of multiple frames. With the distribution-based inter-frame cross constraints, multi-frame poses within the fixed-size window are jointly optimized. Further, this book presents a hierarchical tightly coupled 3D LiDAR-inertial odometry based on distribution, by introducing inertial measurement unit (IMU) to improve the robustness and accuracy of localization. To solve the degradation problem of point cloud distribution constraints in the tight coupling process of IMU and LiDAR, a loss function that dynamically changes with the distribution parameters is designed according to the propagation of point cloud measurement noise in the point-to-distribution distance observation equation. By adjusting the loss term, the anti-degradation point cloud distribution constraints are generated, ensuring the stability of pose estimation under aggressive motions. On this basis, filtering and smoothing are integrated to achieve localization in a hierarchical tightly coupled manner. The low-level constructs joint constraints of IMU prior and anti-degeneration point cloud distribution to estimate the pose of the current frame in real time. The high level combines prior, IMU pre-integration, and point cloud observation constraints to perform fixed-lag smoothing on multi-frame poses. In this way, the efficiency is ensured and a coarse-to-fine odometry estimation with robustness is realized. The effectiveness of the proposed methods is verified through experiments.

Finally, a LiDAR place recognition method is proposed, which is based on range image and cyclic column-shift-invariant attention. The existing range image-based solutions resort to the invariance of NetVLAD for the robustness to point cloud rotation, which restricts the change of scales during feature extraction. To address this problem, a cyclic column shift-invariant attention is designed. It weights outputs of average pooling and maximum pooling on the input feature, which is then cyclically convoluted to generate attention map equivariant to the input. By weighting the input feature based on matrix multiplication, the invariance to the cyclic shift of range image along the column direction is achieved while capturing global contextual information, which provides a prerequisite for the variation of feature scales. Then, a multi-scale feature enhancement module based on spatial downsampling and spatial-channel mixing is presented to mine the information of different scales on the output features of column-shift-invariant (CSI) attention. This enhances the discriminability of the global feature vector and facilitates the place recognition performance. The experiment results verify the effectiveness of the proposed method.

In the future, it is expected that a higher degree of intelligence and adaptability will be integrated into mobile robotic systems. Localization and mapping methods will be investigated further toward the direction of higher accuracy, robustness, and comprehensive perception, by which the mobile robots are capable of executing various tasks under complex environments in a more autonomous way. The involvement of advanced scene representation forms such as 3D Gauss splatting and multi-sensor fusion technology will be applied in the mobile robot fields for long-term and stable task execution.

Index

For Product Safety Concerns and Information please contact our EU
representative GPSR@taylorandfrancis.com
Taylor & Francis Verlag GmbH, Kaufingerstraße 24, 80331 München, Germany

www.ingramcontent.com/pod-product-compliance
Lightning Source LLC
Chambersburg PA
CBHW082006190326
41458CB00010B/3098